SpringerBriefs in Computer Science

Series Editors

Stan Zdonik
Peng Ning
Shashi Shekhar
Jonathan Katz
Xindong Wu
Lakhmi C. Jain
David Padua
Xuemin Shen
Borko Furht
V. S. Subrahmanian
Martial Hebert
Katsushi Ikeuchi
Bruno Siciliano

For further volumes:
http://www.springer.com/series/10028

Reaz Ahmed • Raouf Boutaba

Collaborative Web Hosting

Challenges and Research Directions

 Springer

Reaz Ahmed
David R. Cheriton School
  of Computer Science
University of Waterloo
Waterloo, ON, Canada

Raouf Boutaba
David R. Cheriton School
  of Computer Science
University of Waterloo
Waterloo, ON, Canada

ISSN 2191-5768   ISSN 2191-5776 (electronic)
ISBN 978-3-319-03806-3   ISBN 978-3-319-03807-0 (eBook)
DOI 10.1007/978-3-319-03807-0
Springer Cham Heidelberg New York Dordrecht London

Library of Congress Control Number: 2013955570

Printed on acid-free paper

Springer is part of Springer Science+Business Media (www.springer.com)

Preface

The Web has tremendous importance worldwide. It has arguably become the world's greatest resource for information, and its success has fostered a variety of new ways for people to share information, communicate, and interact. Over the past decade, a wave of cultural phenomena – including Facebook, Google+, Flicker, YouTube, and MySpace – have all utilized the Web as their interface. However, cloud-based solutions for online storage, backup, and sharing of multimedia content over the Web have inherent privacy perils. Users have to put their trust on the cloud-service providers. Service providers dictate the terms of usage, and potentially gain control over users' contents. Beside the privacy concern, transporting huge volumes of user-generated, multimedia content to distant data centers may not be bandwidth friendly for unpopular contents. A peer-to-peer (P2P) Web-based content sharing architecture can subside these problems. This book investigates the challenges in P2P web hosting and presents a potential solution named pWeb. Three major challenges have been addressed in pWeb: (a) persistent naming of Web contents over non-persistent P2P networks, (b) decentralized Web content searching and distributed ranking of search results, and (c) ensuring content availability with minimal replication overhead. pWeb will allow free hosting of websites and multimedia Web contents, without limitation on content type or size. This will provide anybody the opportunity to publish to the masses, rather than restricting them by economics. In addition, freedom of speech is a valued principle; however worldwide there are many who strive to block access to certain information. The distributed approach of pWeb is inherently resistant to censorship, and will help to spread this freedom worldwide.

Waterloo, ON, Canada

Reaz Ahmed
Raouf Boutaba

Contents

Chapter 1
Introduction

Mode of information production, dissemination and consumption is gaining a new momentum with the advent of cheaper and more powerful home entertainment devices (like set-top-boxes, home-gateways, network-attached storages, gaming consoles etc.) and hand-held devices (like smart-phones, tablets, portable gaming devices etc.). Combining the powerful multimedia capabilities of the hand-held devices with the persistent uptime behavior of the home entertainment devices, users can have an elevated Internet experience in a cost-effective manner.

Hand-held devices will increase dramatically in the upcoming years, which is predictable from the prominent shift of the tech industry towards hand-held device market, specially the smart phones and tablet PCs. Equipped with powerful multimedia (e.g., HD video camera, audio, GPS etc.) and networking (e.g., Wi-fi, 4G LTE, Bluetooth etc.) capabilities, these devices are generating voluminous content. These devices are contributing significantly to the popular social networking sites (e.g., Facebook) and online multimedia streaming portals (e.g., YouTube). As of February 2010, YouTube served one billion videos per day, and more interestingly, it would take 35 h to watch the videos uploaded to YouTube per minute. Online storage and backup solutions are yet another class of Internet applications that are consistently gaining popularity. These solutions offer reliable online storage and ease of access over the Internet.

Cloud-based solutions for online storage, backup and sharing of multimedia content over the Web have a few inherent drawbacks as pointed out in [3]. First, voluminous multimedia content has to be uploaded to the cloud-stores, which generates significant amount of Internet traffic. Second, building new data-centers will generate more pressure on the energy sector; as of February, 2009, Microsoft's data center in Quincy, Washington was consuming 48 MW of electricity – sufficient to power around 40,000 homes [1].

Transporting huge volumes of user generated, multimedia content to distant data-centers is not a scalable solution. Rather, semi-persistent devices like, set-top boxes, home-gateways, network-attached storages (NAS) etc., with network and storage capabilities and residing near multimedia content production and consumption

R. Ahmed and R. Boutaba, *Collaborative Web Hosting*, SpringerBriefs in Computer Science, DOI 10.1007/978-3-319-03807-0__1, © The Author(s) 2014

points can be more appropriate targets for placing these contents. This strategy will greatly reduce inter-AS traffic, provide efficient access to delay sensitive multimedia content, and reduce power and resource consumption at the data-centers.

pWeb is a peer-to-peer (P2P) web-hosting infrastructure that will transform networked, home-entertainment devices into light-weight, collaborating Web-servers for persistently storing and serving multimedia and web contents. In pWeb, user generated voluminous multimedia contents will be pro-actively uploaded in a nearby network location (preferably within the same LAN or at least within the same ISP) and a structured P2P mechanism will ensure Internet accessibility by tracking the original contents and their replicas. Clearly, this is a radical departure in how information would be managed compared to the existing Web.

1.1 Importance of P2P Web Hosting

Internet is the largest knowledge base that mankind has ever created. Autonomous hosting infrastructure and voluntary contributions from millions of Internet users have given the Internet its edge. However, contemporary Web search services are governed by centrally controlled search engines, which is not healthy for our online freedom due to the following reasons. A Web search service provider can be compromised to evict certain websites from the search results, which can reduce the websites' visibility. Relative ranking of websites in search results can be biased according to the service providers' preference. Moreover, a service provider can record its users' search history for targeted advertisements or spying. For example, the recent PRISM scandal surfaced the secret role of the major service providers in continuously tracking our web search and browsing history.

A decentralized Web search service can subside these problems by distributing the control over a large number of network nodes. No single authority will control the search result. It will be computed by combining partial results from multiple nodes. Thus a large number of nodes have to be compromised to bias a search result. Moreover, a user's queries will be resolved by different nodes. All of these nodes have to be compromised to accumulate the user's search history.

Distributed indexing and decentralized searching of the Web are very difficult to achieve given the bandwidth limitation and response time constraints. In addition to indexing and searching, a distributed web search engine should be able to rank the search results in a decentralized manner, which requires global knowledge about the hyper-link structure of the Web and keyword-document relevance. Predicting such global information based on local knowledge only is inherently challenging in any large scale distributed system. Moreover, incremental retrieval of search results in a distributed manner is essential for conserving valuable network bandwidth.

Distributed Hash Table (DHT) based systems [6, 8, 10] offer efficient indexing and lookup of information in a distributed manner, yet they does not natively support approximate matching of query keywords to advertised documents. On the other hand, distributed ranking techniques proposed in existing research works [5,7,11] compute approximate ranking of search results based on partial information available locally to each node.

A number of research works [4, 5, 9] and implementations (YacY$^{www.yacy.net}$, Faroo$^{www.faroo.com}$) have focused on distributed Web search and ranking in peer-to-peer (P2P) networks. These approaches have two potential problems in common: (a) *lookup overhead*: number of network messages required for index/peer lookup is much higher in P2P networks compared to a centralized alternative, (b) *churn*: maintaining a consistent index in presence of high peer churn is not feasible. Thus, those solutions have issues with performance and accuracy requirements.

1.2 Challenges

A P2P network is fundamentally different from a client-server architecture. First, peers in a P2P system may join and leave the network frequently, while Web servers are expected to remain up continuously for long periods of time. Second, shared content in a P2P system often moves from one peer to another, whereas web pages do not usually change their location within the Internet. These differences mean that state-of-the-art P2P technology cannot be used directly to create a serverless hosting system. A number of research challenges including the followings must be addressed.

1.2.1 Naming

Web documents are identified using Unified Resource Locators (URLs), which form the hyperlink structure of the World Wide Web. However, URLs are not suitable for naming P2P Web objects, due to peer and content dynamism. The domain name part of a URL essentially specifies the location of a document in the Internet. However, in a P2P environment there is no guarantee of a stable location for a document. Peers' get new network addresses for each session. As a result, the Domain Naming System (DNS), which maps URLs to server IP addresses, is not adequate for naming peer or content in a P2P system. Besides DNS, search engines (like Google or Bing) provide a unanimous mechanism for keyword to content mapping. They crawl the Internet hosts with fixed domain names and DNS resolvable network addresses. In our P2P web scenario, the search engines will not be able to index peers and their contents due to the lack of a proper naming and name resolution scheme. In summary, URL based naming and hierarchical DNS lookup are not suitable for P2P Web hosting.

1.2.2   P2P Web Search

While the current Web relies on centralized search engines (e.g. Google, Yahoo, etc.), traditional P2P search techniques rely on distributed indexing. P2P indexing is done voluntarily by transient peers and there is no central authority for controlling index creation or maintenance. An index is built from important keywords of shared content. However, while this metadata is effective for identifying certain content, web searches rely on full text indexing. Moreover, web search engines support partial keyword matching, which is very hard to achieve in a P2P environment. A successful indexing mechanism for a P2P Web scenario should support full text searching, with partial keyword matching capability. We have previously developed a novel search technique named Plexus [2], that supports efficient partial keyword matching and parallel multiple keyword lookups. Within this project, further investigation will be dedicated to incorporating full text indexing capabilities into the Plexus framework.

Relevance Ranking (RR) (the process of ordering search results by relevance to the search keywords) and IR (the gradual retrieval of search results in parts) are both commonly offered by web search engines. Currently, P2P search techniques return links to all matching contents at once. However, as web search results may return a very large number of (partial) matches, RR and IR techniques are essential to the usability and performance of a P2P Web system. Implementing RR and IR in a distributed manner are challenging problems. While responding to a query, a peer must assess the relevance of its indexed pages, without any global knowledge. For IR, the routing mechanism must also be able to track previously returned results, and already queried peers. This is well beyond the current capabilities of P2P systems and must be developed.

1.2.3   Ensuring Content Availability

In contrast to web servers, the uptime of a typical Internet user is short. In the context of pWeb, it would be required for a peer to remain online round the clock to host its web contents, unless some measure is taken to host the contents during its off-line period. Contemporary P2P techniques rely on content replication to increase availability; however, they do not focus on content persistence over time. Besides the reliability requirement, security and privacy issues have to be considered while placing contents in a P2P web hosting environment.

1.3 Organization

The rest of this book is organized as follows. First, we provide a premier on Plexus in Chapter 2. Then, we address the problems of naming, searching and availability. More specifically, we explain the mechanism for naming peers and contents in Chapter 3. Then we explain in Chapter 4, the mechanisms for Internet compatible web search in pWeb. Finally, we explain the mechanism for improving content availability in Chapter 5. Filnally, we conclude in Chapter 6.

References

1. R. H. Katz, Tech Titans Building Boom, in IEEE Spectrum, Feb. 2009.
2. R. Ahmed and R. Boutaba. Plexus: A scalable peer-to-peer protocol enabling efficient subset search. *IEEE/ACM Transactions on Networking*, 17(1):130–143, February 2009.
3. M. Armbrust, A. Fox, R. Griffith, A. D. Joseph, R. Katz, A. Konwinski, G. Lee, D. Patterson, A. Rabkin, I. Stoica, and M. Zaharia. A view of cloud computing. *Commun. ACM*, 53(4):50–58, Apr. 2010.
4. M. Bender, S. Michel, P. Triantafillou, G. Weikum, and C. Zimmer. P2P content search: Give the web back to the people. In *IPTPS*, 2006.
5. M. Kale and P. S. Thilagam. DYNA-RANK: Efficient Calculation and Updation of PageRank. In *ICCSIT*, pages 808–812, 2008.
6. P. Maymounkov and D. Mazireres. Kademlia: A Peer-to-Peer information system based on the XOR metric. pages 53–65. Springer-Verlag, march 2002.
7. J. X. Parreira and G. Weikum. JXP: Global Authority Scores in a P2P Network. In *8th Int. Workshop on Web and Databases (WebDB)*, 2005.
8. S. Ratnasamy, P. Francis, M. Handley, R. Karp, and S. Schenker. A scalable content-addressable network. pages 161–172, 2001.
9. K. Sankaralingam, M. Yalamanchi, S. Sethumadhavan, and J. Browne. Pagerank computation and keyword search on distributed systems and p2p networks. *Journal of Grid Comp.*, 1(3):291–307, 2003.
10. I. Stoica, R. Morris, D. Liben-Nowell, D. R. Karger, M. F. Kaashoek, F. Dabek, and H. Balakrishnan. Chord: a scalable Peer-to-Peer lookup protocol for Internet applications. *IEEE/ACM Transaction on Networking (TON)*, 11(1):17–32, 2003.
11. J. Wu and K. Aberer. Using a Layered Markov Model for Distributed Web Ranking Computation. In *Proc. ICDCS*, pages 533–542, Jun. 2005.

Chapter 2
Plexus: Routing and Indexing

A scalable and distributed indexing mechanism is essential for any P2P web hosting solution. We also need to search indexed information in a bandwidth efficient manner. Search and indexing is essential for a number of reasons. First, we need to maintain information about each peer in the system, e.g., their name and IP address binding, availability information, last seen time etc. Second, we need to keep track of the Web contents hosted in different peers. Third, we need to keep track of the relative importance or popularity of the available contents for ranking them in the search results. We have previously developed a distributed search technique named Plexus [1] that supports bandwidth efficient search and approximate matching. Plexus has been intensively used for pWeb deployment. In this section, we present the basic protocols in Plexus. We also outline a few enhancements to the basic Plexus protocol to support the above mentioned requirements.

While the Internet relies on dedicated search engines (e.g., Google, Yahoo!, Bing etc.), traditional P2P search techniques rely on distributed indexing. P2P indexing is done voluntarily by transient peers and there is no central authority for controlling index creation or maintenance. An index is built from important keywords of a shared content. Although metadata is effective for identifying certain content, web searches rely on full text indexing. Moreover, web search engines support partial keyword matching, which is very hard to achieve in a P2P environment without sacrificing routing efficiency. A successful indexing mechanism for the pWeb scenario should support full text search, with partial keyword matching capability.

Like other DHT techniques Plexus supports efficient routing, which scales logarithmically with network size. In addition, support for approximate matching is built into the Plexus routing mechanism, which is not easily achievable by other DHT techniques. To cope with churn in P2P systems, Plexus supports multipath routing and efficient replica placement. Plexus delivers a high level of fault-resilience by using replication and redundant routing paths. Because of these advantages, we have incorporated Plexus routing at the core of our P2P web hosting solution.

R. Ahmed and R. Boutaba, *Collaborative Web Hosting*, SpringerBriefs in Computer Science, DOI 10.1007/978-3-319-03807-0_2, © The Author(s) 2014

2.1 Core Concepts in Plexus

In Plexus, a search or advertisement keyword is converted to a Bloom filter [2], which is a compact data-structure used to represent a set. However, the set membership test operation may result into false (erroneously) positives with a small probability. An m-bit array is used to represent a Bloom filter. $\hbar$ different hash functions need also to be defined. In an empty Bloom filter all the bits are set to zero. To insert an element in a Bloom filter, it is hashed with the $\hbar$ hash functions to obtain $\hbar$ positions in the bit-array and corresponding $\hbar$ bits are set to 1. The membership test process is similar to the insert process. The element, say x, to be tested for set membership, is hashed with the same $\hbar$ hash functions and corresponding $\hbar$ positions in the bit-array are checked. If any of these $\hbar$ bits is not 1 then x is definitely not a member of the set represented by this Bloom filter. On the other hand, if all of the $\hbar$ bits equal to 1, then there is a high probability that x is a member of the set.

A Hamming distance based technique derived from the theory of *Linear Covering Codes* [3] is used for routing. The keyword to Bloom filter mapping process retains the notion of similarity between keywords, while Hamming distance based routing delivers deterministic results and efficient bandwidth usage.

In Plexus, advertisements and queries are routed to two different sets of peers in such a way that the queried set of peers and the advertised set of peers have at least one peer in common, whenever a query pattern is within a pre-specified Hamming distance of an advertised pattern. As explained in Fig. 2.1, a linear covering code ($\mathscr{C}$) partitions the entire pattern space $\mathbb{F}_2^n$ into Hamming spheres, represented by hexagons. A codeword ($c_i \in \mathscr{C}$) is selected as the unique representative for all the patterns within its Hamming sphere. To facilitate approximate matching in Plexus, an advertisement pattern, say P; is mapped to all codewords, denoted by $\mathscr{A}(P)$, that are within a pre-specified Hamming distance, say s, from P. Mathematically $\mathscr{A}(P)$ can re represented as, $\mathscr{A}(P) = B_s(P) \cap \mathscr{C} = \{Y | Y \in \mathscr{C} \wedge d(Y,P) \leq s\}$, where $B_s(P)$ is the Hamming sphere of radius s centred at P and $d(Y,P) = |Y \oplus P|$ is the Hamming distance between Y and P. Similarly, a query pattern, say Q, is mapped to

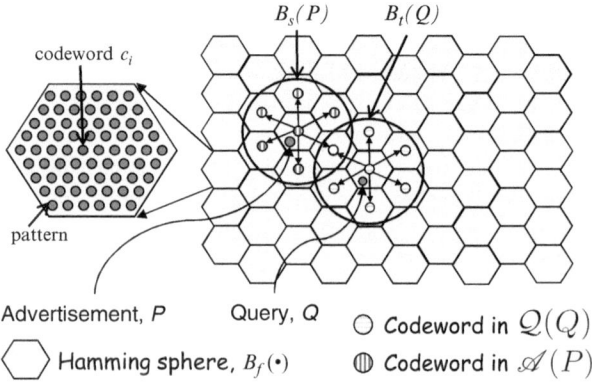

Fig. 2.1 Hamming distance based indexing

a set of codewords $\mathscr{Q}(Q) = B_t(Q) \cap \mathscr{C}$, for some pre-specified Hamming distance t. It is shown in [1] that there will be at-least one common codeword in $\mathscr{A}(P)$ and $\mathscr{Q}(Q)$, if $d(P,Q) \leq s+t-2f$, where f is the covering radius of $\mathscr{C}$. In other words, by looking into the codewords in $\mathscr{Q}(Q)$, one should be able to find all advertised patterns within Hamming distance $s+t-2f$ from Q.

2.2 Plexus Routing

Consider a (n,k,d) linear covering code $\mathscr{C}$ with generator matrix $G_{\mathscr{C}} = [g_1, g_2, \ldots, g_k]^T$. To route using this code, a peer responsible for codeword X, has to maintain links to $(k+1)$ peers with codewords $X_1, X_2, \ldots, X_{k+1}$, computed as follows:

$$X_i = \begin{cases} X \oplus g_i & 1 \leq i \leq k \\ X \oplus g_1 \oplus g_2 \oplus \ldots \oplus g_k & i = k+1 \end{cases} \quad (2.1)$$

Now, the routing process in Plexus can be best explained by the example in Fig. 2.2, which shows the possible routes from peer X to peer $Y = g_2 \oplus g_3 \oplus g_3$ (any codeword Y can be generated from any other codeword X as follows: $Y = (X \oplus g_{i_1} \oplus g_{i_2} \oplus \ldots \oplus g_{i_t})$, where $g_{i_1}, g_{i_2}, \ldots g_{i_t} \in G$ and $\oplus$ is bitwise XOR operation). Peer X will forward the message to any of $X_2(= X \oplus g_2)$, $X_3(= X \oplus g_3)$ or $X_5(= X \oplus g_5)$, which are one hop nearer to Y than X. If the message is forwarded to X_2 then X_2 can route the message to Y via $X_{23}(= X \oplus g_2 \oplus g_3)$ or $X_{25}(= X \oplus g_2 \oplus g_5)$. In such an overlay, it is possible to route a query from any source to any destination codeword in $\frac{k}{2}$ or fewer routing hops [1].

In Plexus protocol, a peer say Y replicates its indices to peer Y_{K+1}. In presence of failure a peer's replica can be reached in just two extra hops, which can be explained using the example of Fig. 2.3. Here peer X is attempting to route a query to peer Y, which has failed. When a neighbor (Y') of Y detects the failure, it forwards the query to its own replica Y'_{K+1} in one hop. Next peer Y'_{K+1} forward the query to peer Y's replica Y_{K+1} in one hop.

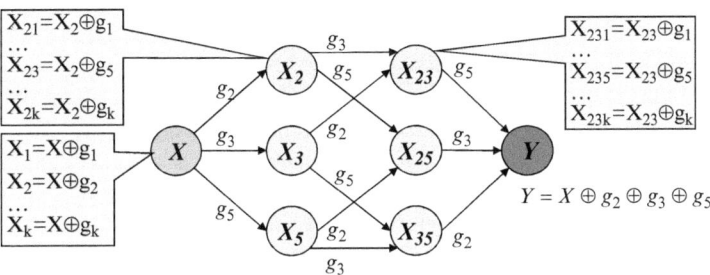

Fig. 2.2 Possible routing paths between peer X and Y

Fig. 2.3 Routing under
failure

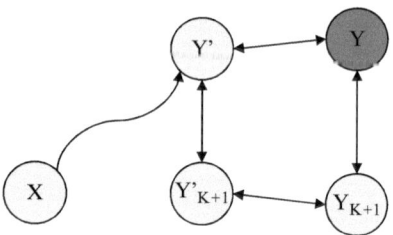

2.3   Using Plexus in pWeb

Plexus is optimized for P2P content sharing environments, which has several behavioral differences compared to the P2P Web hosting scenario:

- *Replication behavior*: In a P2P network, downloaded copy of a shared content becomes a source for future downloads. While in our context, authenticity is an important factor governing a web content's placement. A popular content may be replicated at multiple locations, but content authenticity has to be ensured.
- *Query behavior*: The number and variety of documents in the P2P Web scenario will be much higher than that in a P2P content sharing system. This will result into higher query traffic and index volume.
- *User connectivity pattern*: In pWeb, peers are expected to host web contents for longer periods of time, compared to the peers in a traditional P2P content sharing system. Peers will join and leave the network periodically, but it is expected that a returning peer will retain the replicated contents from its previous session and will continue to host those contents.
- *Full text indexing*: In content sharing P2P systems, a few keywords are advertised for each shared content, while a matching query string comprises a subset of the advertised keywords. On the contrary, Web search engines use many important keywords per webpage, while web queries involve a few keywords. In essence the gap between the number of keywords per advertisement and the number of keywords per query is much higher in Web scenario compared to P2P content sharing scenario.

The original Plexus routing mechanism has to be modified in order to handle the above mentioned behavioral differences between P2P web hosting and P2P content sharing. We utilize the inherent capability of Plexus for trading off query traffic with advertisement traffic. As the expected advertisement rate in our case is much smaller than query rate, we can increase the number of nodes indexing a content, which will help in reducing the number of nodes to be searched for query lookup. To cope with ad hoc connectivity in the peers, we will assign each peer a unique name. This will enable a peer to host websites or contents from its previous sessions. Differential updates will be propagated to the returning peer during the rejoin process.

To incorporate full text indexing capability, we have modified the advertisement mechanism in Plexus framework [1] as follows (see Fig. 2.4).

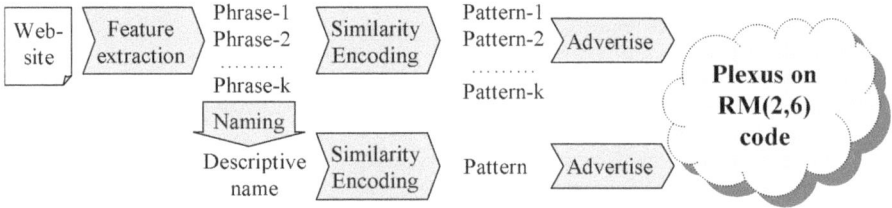

Fig. 2.4 Modified Plexus framework for web document advertisement

- Instead of advertising one pattern per document we advertise one pattern per phrase. Phrases are extracted by applying a feature extraction mechanism, such as Latent Semantic Indexing (LSI), over all the advertised web content on a given website. This enables us to perform keyword search on the webpages, in a similar way we search the Internet.
- For each extracted phrase we will first apply a phonetic algorithm like Soundex or Metaphone, and hash the resulting words into a Bloom Filter. There is one Bloom filter per phrase. Use of phonetic encoding increases the degree of similarity matching offered by Plexus.
- Since the expected edit distance between the advertised phrases and query keywords is small, we use a second order Reed-Muller code, instead of the Extended Golay code G_{24}, as proposed in the original Plexus protocol.

In the following chapters, we explain each of these concepts in further detail. More specifically, we explain the mechanism for naming peers and contents in Chap. 3. Then we explain in Chap. 4, the mechanisms for Internet compatible web search in pWeb. Finally, we explain the mechanism for improving content availability in Chap. 5. Each of these mechanisms use Plexus as explain in this chapter.

References

1. R. Ahmed and R. Boutaba. Plexus: A scalable peer-to-peer protocol enabling efficient subset search. *IEEE/ACM Transactions on Networking*, 17(1):130–143, February 2009.
2. B. H. Bloom. Space/time trade-offs in hash coding with allowable errors. *Communications of ACM*, 13(7):422–426, 1970.
3. G. Cohen. *Covering codes*, volume 54. North Holland, 1997.

Chapter 3
Naming

Web documents are identified using Unified Resource Locators (URLs), which form the hyperlink structure of the World Wide Web. However, URLs and the contemporary Domain Naming System (DNS) resolution may not be suitable for naming P2P Web objects, due to peer and content dynamism. The domain name part of a URL essentially specifies the location of a document in the Internet. However, in a P2P environment there is no guarantee of a stable location for a document. A DNS resolver maps URLs to server IP addresses, which allows site relocation and replication without affecting the URL. Site relocation is relatively less frequent than P2P content dynamism. DNS updates are almost static compared to P2P index updates, which allows DNS clients to cache network addresses for saving network bandwidth and improving response time. Hence, URL based naming and hierarchical DNS lookup may not be used in its current state for P2P Web hosting.

3.1 Requirements

A suitable naming system for P2P Web deployment should be independent of the spatial and temporal scope of the referred document. There should also exist easy conversion mechanism for converting URLs to the new naming system, and vice versa. P2P Web system requires a human-readable, flexible naming scheme. The naming authority should be distributed as well as the name resolution architecture. The naming scheme should be compatible with widely accepted Internet naming standards. Below is a list of requirements for naming Web contents in a P2P Web scenario:

- *Readability*: Names may or may not be considered *human-readable*. Readable names are important in the P2P Web scenario to facilitate memorization and usability.

R. Ahmed and R. Boutaba, *Collaborative Web Hosting*, SpringerBriefs in Computer Science, DOI 10.1007/978-3-319-03807-0_3, © The Author(s) 2014

- *Extensibility*: The naming system should be extensible for future updates. Updates can occur in the format and structure, in the scope, and in the grammar. In a rapidly evolving P2P environment, an extensible naming scheme is desirable. New names should remain compatible with existing, older names from the same naming schemes. The naming scheme should be able to accommodate changes in namespace and scope.
- *Namespace size*: Namespace size determines how many unique entities can be named. Although a finite namespace allows a fixed allocation for storing and transmitting names, it runs the risk of being exhausted. For a P2P environment, an infinite namespace is ideal.
- *Naming authority*: A naming authority is responsible for assigning and changing names, as well as preventing name conflicts. It can be centralized or decentralized. For the centralized case a single entity is responsible for the entire namespace, while for the distributed case the namespace is divided in non-overlapping domains and each domain is managed by a separate authority. In a P2P Web context, distribution is preferable to avoid performance bottlenecks and a single point of failure.
- *Name resolution architecture*: The name resolution process determines how to translate names to addresses. Contrary to the existing name resolution systems that return at most one address for a name, pWeb requires a system that will return a set of peer addresses. The name resolution architecture should also be distributed.
- *Name persistence*: A *static* name permanently denotes the same object, while a *dynamic* name is assigned to an object for the lifetime of that object. Location dependent static names (e.g., URLs) are used for naming Internet documents. However, they are not suitable for naming documents in P2P environments due to the lack of a central naming authority and the high level of dynamism.
- *Standardization and implementation*: A naming scheme based on a well-defined standard, which is easy to incorporate into the existing infrastructure is far easier to deploy than an experimental, constantly-changing scheme, even if the latter is superior in other aspects. Ideally, the naming scheme used in a P2P Web hosting system should comply with widely accepted standards and should have most components of its name resolution already in place.

In summary, P2P Web system requires a human-readable, flexible naming scheme that is persistent and location independent. The naming authority should be distributed as well as the name resolution architecture. The naming scheme should be compatible with widely accepted Internet naming standards.

3.2   Who Needs a Name

Facilitating a persistent naming scheme on a non-persistent, transient P2P network is a challenging problem. To achieve this goal, we propose a multi-faced naming scheme, named pRL (P2P Resource Locater), which comprises of three components: (i) a UUID for system use, (ii) a human-friendly component and (iii) a set of descriptive key-value pairs. Within the pWeb framework, we will need to name the following four entities:

1. **Peer:** Peer names should be unique within pWeb system. A peer has to register a name with the system before using it. A returning peer should reclaim its registered name using a challenge/response mechanism.
2. **Group:** Websites will be replicated within well defined, automatically maintained small groups of peers. These groups will not be visible to the users. Hence group names will only have the UUID part of a name.
3. **Website:** To separate a website from the hosting peer, we will use separate namespaces for websites and peers. The originating peer's name and the replication group's UUID will be stored in the web-site's key value list. We will allow multiple website to have same name, but the UUID and key-value list will be different. For disambiguating between multiple names, the system will use the UUID part and the users will use the key-value list.
4. **Web pages:** For naming webpages or documents, we intend to use hierarchical path names relative to the website's pRL, much like relative names in URL scheme.

3.3   Naming in Peer-to-Peer Systems

In this section we highlight the naming schemes and the name resolution systems in three categories of P2P networks: (a) file-sharing systems, (b) BitTorrent and (c) P2P DNS.

3.3.1   File Sharing Systems

Content sharing P2P systems (e.g., Gnutella, Kaaza, Morpheus etc.) use descriptive keyword list for content naming. Those systems use randomly selected, temporary identifiers for peer names. In this approach content names are unique and hence can not be used as a substitute for hyperlinks. On the other hand, peers do not retain their IDs across sessions, thus peer history from the previous sessions can not be reused in the indexing process.

3.3.2   BitTorrent

BitTorrent [2] is a P2P application that replicates content on multiple peers that are interested in that content. A specificity of BitTorrent is the notion of torrent, which

BitTorrent download info

- tracker version: 3.2.2
- server time: 2003-07-14 15:17 UTC

info hash	complete	downloading	downloaded
01fb5fcd21b4f6fc7fbbe6b812e4bffe08a3edfc	0	3	0
041c08e1a009bfa8c9be7117d5f0372ec68dcdbd	0	6	15
162f5bba51dac70ae28433031612ae1b0be2dfe4	1	25	273
1aeb2d925c325662321e67a07a36a60d0876f3f7	3	10	336
| | | | | | | | | |	|	|	|
| | | | | | | | | |	|	|	|
| | | | | | | | | |	|	|	|
e1d9efefc450f7af6a2b56038335699e1a2786b0	9	43	833
f29bc2004c0eb013608c59469a0fd899baa434ea	0	13	138
fdd4dfda29477ad065bb4d6478a01019b4358268	6	36	840
0 files	86/97	480/649	10308/12200

- *info hash:* SHA1 hash of the "info" section of the metainfo (*.torrent)
- *complete:* number of connected clients with the complete file (total: unique IPs/total connections)
- *downloading:* number of connected clients still downloading (total: unique IPs/total connections)
- *downloaded:* reported complete downloads (total: current/all)
- *transferred:* torrent size * total downloaded (does not include partial transfers)

Fig. 3.1 A sample torrent file

defines a session of transfer of a single content to a set of peers. Peers involved in a torrent cooperate to replicate the file among each other using swarming techniques. A user joins an existing torrent by downloading a *.torrent* file, usually from a Web server that contains metadata, e.g., the piece size and the SHA-1 hash values of each piece, and the IP address of the so-called tracker of the torrent. The tracker is the only centralized component of BitTorrent, but it is not involved in the actual distribution of the file. It keeps track of the peers currently involved in the torrent and collects statistics on the torrent. A sample torrent file is presented in Fig. 3.1.

The peers in a BitTorrent network are assigned random IDs. A torrent file, on the other hand, contains SHA-1 hash of each piece of the target file. These hash values can be used to uniquely identify the pieces of a file. Its use of SHA-1 hash in identifying pieces of a file introduced a new idea in the realm of P2P naming.

3.3.3   P2P DNS

A number of research works, including [3, 6] and [8], focus on implementing DNS lookup using P2P systems. We briefly explain each of these approaches.

In [3], the authors have proposed a distributed hash mechanism called *DHash* – a Chord-based hash table. DHash improves over the load balancing and robustness properties of Chord's consistent hashing technique. DNS records are cached along every node on a lookup according to the Chord routing protocol. To improve robustness, a fixed number of replicas are maintained for each name record. DHash also uses public key cryptography for ensuring name security.

Cooperative Domain Name System (CoDoNS) [6] focuses on fast DNS lookup, resilience to denial of service (DoS) attack and low update propagation delay. CoDoNS is a legacy DNS compatible naming system. It uses Beehive [7] DHT to proactively replicate a DNS record according to its request rate. Replica placement and synchronization overheads in CoDoNS is very high. It increases exponentially with the popularity of a DNS record.

Internet Indirection Infrastructure (*i3*) [8] is a general-purpose framework for rendezvous-based communication services like unicast, multicast, anycast and mobility. *i3* is built on top of Chord routing. *i3* can be used for DNS record indexing and lookup. However, the expected lookup latency in *i3* is greater than that of the legacy DNS.

All of these works use DHT-techniques for P2P lookup and support exact name lookup. Updating name to IP association over DHT lookup is expensive in terms of bandwidth. Besides, the achieved response time for name resolution would be significantly higher than that of the legacy DNS.

3.4 A Collaborative Naming Scheme

For P2P Web hosting we need a persistent, secure and human friendly naming scheme. The naming scheme should have flexible embedding of organizational structure to permit seamless content movement between the peers. The name to IP binding should be verifiable from the peer that requested the name resolution. We proposed a naming scheme and a name resolution mechanism to achieve all of these properties in [1]. In this section we provide a brief overview of that work.

A significant difference of the pWeb naming scheme from other P2P naming schemes is the assignment of names to content instead of peers. Here names are assigned to websites, not to peers. We coin the term pRL which stands for pWeb Resource Locator. A pRL is used in pWeb to uniquely identify any resource just as an URL used in the Internet. However peers are identified by the underlying Plexus codewords. A peer can have multiple websites hosted on it and thus have multiple pRLs resolved to its IP address and port number. A pRL uniquely identifies a website but the reverse is not true. A single website can have multiple pRLs associated with it. This feature can also be used for load balancing in case of hot spots. In the following sections we first describe the entities to be named along with the requirements, the naming framework and then we discuss the naming authority and name resolution systems. After that we present two naming techniques that we devised for pWeb.

3.4.1 Entities and Requirements

In this section we discuss the entities that need to be named and the naming requirements for each entity. In order to facilitate a persistent object naming scheme on a non-persistent, transient P2P network, we need to name the following entities in the pWeb system:

- **Website Naming:** As stated earlier, pRLs are assigned to websites and used just like URLs in the Internet. The process of pRL assignment can be fully automated. We also have the provision for assigning human friendly names to a website. This is implemented as an additional level of mapping in the name resolution system. As pRLs are assigned to the content instead of the peer, name persistence is supported across peer sessions and any peer with a valid copy (replica) of the content can serve as a source. This makes it possible to provide support for peer data browsing, bookmarking, IM etc. between peers.
- **Content Naming:** For naming the content of a website we use relative naming and the information is contained in the Object ID (Fig. 3.5) field of pRL. Content metadata (e.g., descriptive keywords) are associated with each object/content in pWeb. This facilitates keyword searching.
- **Peer Naming:** A peer is not assigned any name under the pWeb naming scheme. However, Plexus assigns a codeword to each peer in the pWeb overlay network, responsible for a specific codeword. This codeword serves as the peer ID.
- **Replication Group Naming:** To ensure document persistence, we have to first ensure a persistent storage over the non-persistent P2P network. To this end, we devised a novel replication mechanism that is expected to work across temporal and spatial dimensions. This replication mechanism is based on the diurnal availability pattern of the peers participating from different time zones across the Globe. Peers from different time zones form a group for hosting a web site. In effect, peers in a group host a website in turn to make it available round the clock.

  From naming point of view a group should have globally unique, auto-generated ID. We do not require the Group IDs to be human-friendly, since these IDs are expected to be used internally by the name resolution mechanism for locating the currently active (i.e., alive) replica of a web site.

  Although replication Group ID will introduce an additional level of indirection during the name resolution process, it will decouple the location of a content from its name.

3.4.2 pWeb Naming System

The pWeb system architecture and the functional dependencies within the architectural components have a direct impact on the naming scheme and vice versa. Hence for the sake of a better understanding of the naming scheme,

Fig. 3.2 Advertisement of
a site

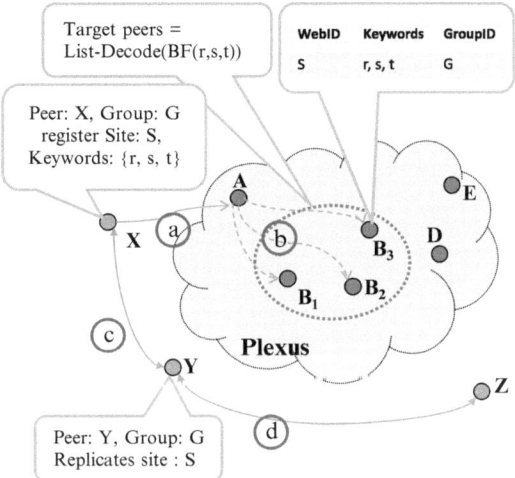

we present the functional architecture of the following three processes within
pWeb: (a) advertisement process, (b) peer rejoin process and (c) query process.

3.4.2.1 Advertisement

To facilitate efficient search of web sites we use a distributed Hamming distance
based indexing mechanism using Plexus routing. As depicted in Fig. 3.2, the
advertisement process consists of the following four steps:

- Step a: Each peer in the system will belong to a replication group. Suppose peer
  X belongs to group G and wants to advertise site S. Assume that the search
  keywords (or other meta information) related to site S are r, s and t. Peer X sends
  this information to a peer A in the pWeb indexing framework.
- Step b: in this step, peer A creates an advertisement pattern (basically a Bloom
  Filter) from this meta information (i.e., r, s and t), list decodes the pattern
  and computes the set of codewords within a pre-specified Hamming distance
  from the advertised pattern. Then it uses the Plexus routing mechanism to
  multi-cast the site index to the peers (B_1, B_2, B_3) responsible for the codewords. The indices stored in the indexing peers (i.e., B_i's) are in the form of
  $<$webID,Keywords,GroupID$>$ triplets.
- Steps c and d: Newly hosted sites or updates in existing sites are propagated to all
  the members (i.e., peer Y and peer Z) of the hosting peer's (i.e., X's) replication
  group (i.e., group G). This replication takes place whenever a group member
  rejoins the network. More detail on the rejoin and replication process is given in
  the next section.

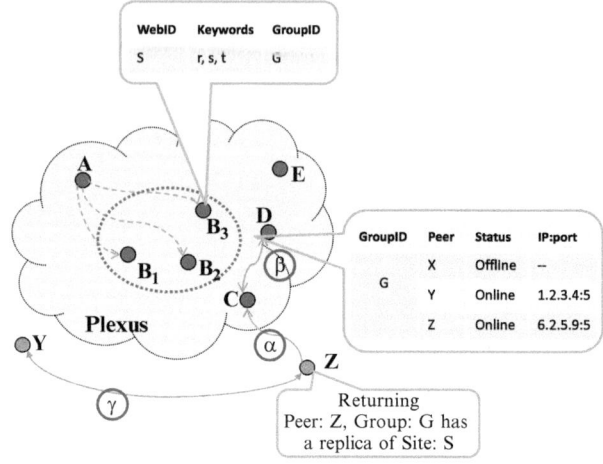

Fig. 3.3 Accommodating returning peer

3.4.2.2   Rejoin and Group Maintenance

In order to maintain diurnal availability, web site hosting peers collaborate in small groups in such a way that at any given instance at least one peer from a group is online with very high probability. Contents in each peer of a group are fully replicated and synchronized. To ensure replica consistency, whenever a peer returns to the system it finds its group members and updates its online status in the following manner:

- Step α: As depicted in Fig. 3.3, peer Z, a member of group G, becomes online after being offline for a while. After becoming online, peer Z request a peer, say C, in the pWeb indexing framework to find the members in its own group G.
- Step β: Peer C, constructs a pattern from the group ID G, decodes the pattern to find the closest codeword, and routes the query to the responsible peer (here D) using Plexus routing. Upon receiving the query, peer D updates the current status of peer Z to online, records its IP:port and returns the IP:port list of all the online members of group G.
- Step γ: In this manner peer Z learns about the currently online members of its own group and synchronizes each others replicas.

3.4.2.3   Query

pWeb can handle two types of queries: (a) keyword search and (b) pRL search. The second case is more straightforward and a subset of the first case. Hence we explain here the first case only, i.e. query by keywords. As depicted in Fig. 3.4, the keyword search process can be performed in the following four steps:

Fig. 3.4 Keyword-based
content searching

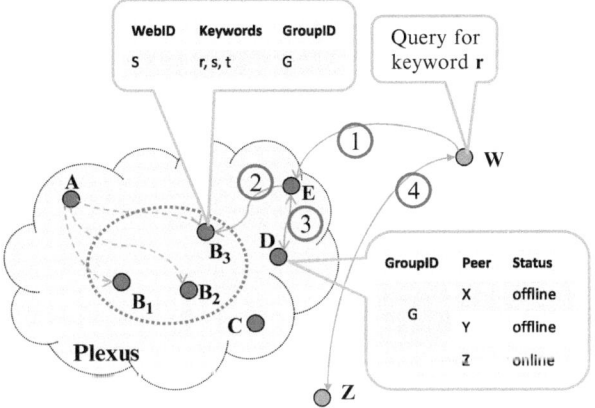

- Step 1: In the example scenario peer W is searching for the sites that have
  keyword r. It first sends the query to a peer, say E, participating in the pWeb
  indexing framework.
- Step 2: Upon receiving the query, peer E constructs a query pattern (a Bloom
  filter) from the query keywords and uses list decoding to find the codewords
  within a pre-specified Hamming distance. Then it uses the Plexus routing to
  forward the query to the peers that are likely to have the meta-information on
  sites with the queried keywords. In the example, peer B_3 responds with the site
  name, keywords and the group ID (G) of the group hosting the site.
- Step 3: Once peer E receives the group ID G, it query the Plexus network (similar
  to the rejoin process) to find the list of currently online members of group G.
  In this instance the IP:Port of peer Z is discovered and returned to the original
  querying peer W.
- Step 4: Now peer W can directly browse site S from peer Z using the pWeb
  hosting framework.

3.4.3   Naming Scheme

In pWeb, each website is assigned a unique ID called pRL. pRLs are globally unique
and maintained by the naming authority described in the next section. Figure 3.5
shows the structure of a pRL and the accompanying metadata.

The pRL consists of three parts: NSID, Web ID and Object ID. NSID identifies
the naming scheme in use. Naming authority and resolution system use this
information for determining the appropriate name resolution method. Using NSID
we can incorporate additional naming scheme into our system with minimum
effort. As P2P web applications consist of a diverse communities of users, we kept
this option open to tackle future architectural changes. The Web ID is a globally

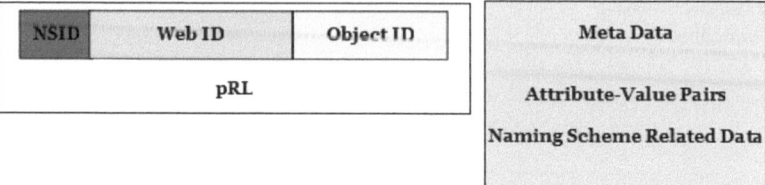

Fig. 3.5 Structure of pRL

unique ID that is associated with a user's web site/domain (we will use the terms web site and domain interchangeably). There must be one such ID per web site. We will discuss more about Web ID in the following sections. The Object ID is used for identifying pages, image, documents etc. under a web site and it is the publisher/owner's responsibility to make the names unique under his/her domain. This requirement is similar to that of any traditional web site.

The metadata is an XML document that contains additional information about the website like title, owner name, description, content, keywords etc. The metadata will facilitate attribute based keyword search in pWeb. The metadata also includes security related information like public key, encryption method etc.

3.4.4 Naming Authority

The primary function of a Naming Authority (NA) is to maintain name uniqueness. A centralized NA will make the task fairly simple, but this design choice is not suitable for pWeb for scalability concerns. Instead, we propose a distributed Naming Authority. Each peer is responsible for maintaining only a portion of the namespace. Responsibility is determined by the *encode(pattern)* function that performs an exact decoding to the nearest codeword from the given *pattern*. The API of the naming authority includes the following functions: RegisterWebID(WebID) and InvalidateWebID(WebID). These two functions are formally presented in Algorithms 1 and 3, respectively.

The function *RegisterWebID(webID)*, is used by a peer to register its Web IDs with the naming authority. A peer first generates a Web ID for its website. The details about the Web ID generation process will be discussed in Sect. 3.4.6. The function first finds out the primary code word (*targetCodeWord*) responsible for the Web ID. Then it creates a list of code words that fall within the Hamming sphere of radius r (which is a global parameter) centered at *targetCodeWord*. The function iterates over the list and registers the Web ID by checking whether the peer associated with the code word is alive or not and then checking the uniqueness of the Web ID using the API function *IsPeerAlive* and *Register*. *Register* simply saves the mapping between pRL and peer IP:port address. If the function can successfully register the Web ID at least at one peer then it returns true otherwise returns false. If failure occurs then the peer has to generate another Web ID and try to register again.

Algorithm 1 RegisterWebID(webID)
_____
1: *count* ← 0
2: *targetCodeWord* ← *encode*(*webId*)
3: *codeWordList* ← [*targetCodeWord*, *decode*(*targetCodeWord*, *r*)] {*r* is a global parameter}
4: **for all** *codeWord* ∈ *codeWordList* **do**
5:     **if** *IsPeerAlive*(*codeWord*) = true **then**
6:         **if** *Register*(*codeWord*, *webID*) = true **then**
7:             *count* ← *count* + 1
8:         **end if**
9:     **end if**
10: **end for**
11: **if** *count* > 0 **then**
12:     *return* true
13: **else**
14:     return false
15: **end if**
_____

The peer associated with the primary code word is responsible for keeping the pRL mapping data replicated across the overlay network. Every peer storing some pRL mapping runs Algorithm 2 to maintain a suitable number of replicas. The analysis for quantifying the suitable number of replicas in diverse network conditions is determined by the Plexus routing protocol.

Algorithm 2 ReplicateWebID(webID)
_____
1: *webIDList* ← WebIDs from locally stored pRL mappings
2: **for all** *webID* ∈ *webIDList* **do**
3:     *targetCodeWord* ← *encode*(*webId*)
4:     **if** *targetCodeWord* = *localCodeWord* **then**
5:         *codeWordList* ← [*decode*(*targetCodeWord*, *r*)]
6:         **for all** *codeWord* ∈ *codeWordList* **do**
7:             *Register*(*codeWord*, *webID*)
8:         **end for**
9:     **end if**
10: **end for**
_____

When a peer decides to delete its website from pWeb, it uses the function *InvalidateWebID*(*WebID*). This function invalidates all pRL mapping replicas from all peers.

Algorithm 3 InvalidateWebID(webID)
_____
1: *targetCodeWord* ← *encode*(*webId*)
2: *codeWordList* ← [*targetCodeWord*, *decode*(*targetCodeWord*, *r*)] {*r* is a global parameter}
3: **for all** *codeWord* ∈ *codeWordList* **do**
4:     *Invalidate*(*codeWord*, *webID*)
5: **end for**
_____

3.4.5   Name Resolution

The name resolution system is pretty straight forward and follows the same principle as the algorithms presented in the previous section. When a peer wants to resolve a pRL, it uses the *encode* function to find out the primary codeword, and the *decode* function to find the codewords within Hamming radius (r) of the primary codeword (i.e., *targetCodeWord*) . Then the peers associated with the code words are queried for the required pRL mapping. This procedure is formally presented in Algorithm 4.

Algorithm 4 Resolve(webID)

1: $targetCodeWord \leftarrow encode(webId)$
2: $codeWordList \leftarrow [targetCodeWord, decode(targetCodeWord, r)]$ {r is a global parameter}
3: **for all** $codeWord \in codeWordList$ **do**
4:     $PeerIP : Port \leftarrow GetMapping(codeWord, webID)$
5:     **if** $PeerIP : Port != NULL$ **then**
6:         return $PeerIP : Port$
7:     **end if**
8: **end for**

3.4.6   Methods for Selecting Web ID

In this section we describe two methods for selecting Web IDs. The first method is based on Public/Private key based naming similar to DONA [5] and NetInf [4]. The second approach is an email address based registration scheme. While the first approach provides us security features like content authentication and data integrity, the second approach offers us no security. However the second approach is easier to implement and incurs lesser computation overhead. We intend to use the Public/Private key based approach when higher level of trust worthiness is required and use the second scheme in other cases. Trust worthiness can also be achieved in the second approach using techniques like community based reputation scores. In the following sections we discuss these two naming schemes in greater detail.

3.4.6.1   Public/Private Key Based Naming

When a peer deploys a new website the pWeb client application creates a pair of Public and Private key, and uses the hash (e.g. SHA1) of the Public key as the Web ID for the website. Using the Private key the peer creates a signature of every web page and includes the signature in the accompanying metadata. The metadata also contains the full Public key of the web page along with other attribute–value pairs for facilitating keyword based search. A receiver hashes the public key in the metadata to see whether it matches the hash in the pRL. Then the receiver decrypts the signature using the public key. If the decrypted string matches the hash of the entire file then the data is both authentic and there was no unauthorized changes in the data.

We have to ensure two things for this scheme to work. Whenever a peer updates its website its signature needs to be updated. And second, a receiver has to hash and decrypt the received content to ensure data authenticity and integrity. The computational overhead of these functions is not so significant for even a nominal personal computer [4].

3.4.6.2 Email Based Naming

In this naming scheme we follow the traditional scheme of registering in any website. When we want to register at a website we need to provide an email address along with a password. Later a verification link is emailed to us for registering at that web site. We follow the same paradigm here. When a peer needs to generate a Web ID it simply uses a user supplied email address as the Web ID and calls the *RegisterWebID* function. The primary code word holder (the peer responsible for registering this Web ID according to Algorithm 1) registers the Web Id and sends back a verification link (pRL link) back to the users inbox. The user finishes the registration process by browsing to the pRL. The choice of email based naming is motivated by the need for avoiding Web ID name conflicts and for preventing malicious nodes from launching DoS attacks.

3.5 Summary

In this chapter, we have identified the naming requirements for a P2P web hosting system. We also identified the entities that has to be named for P2P web hosting. We presented a few example scenarios describing advertisement, rejoin and query processes withing the pWeb architecture. Finally, we have outlined two alternative algorithms for achieving unique and secure naming in pWeb.

References

1. M. Bari, M. Haque, R. Ahmed, R. Boutaba, and B. Mathieu. A naming scheme for p2p web hosting. *Selected Areas in Communications, IEEE Journal on*, 31(9):299–309, 2013.
2. B. Cohen. Incentives build robustness in BitTorrent. In *Workshop on Economics of Peer-to-Peer systems*, volume 6, pages 68–72. Citeseer, 2003.
3. R. Cox, A. Muthitacharoen, and R. T. Morris. Serving dns using a peer-to-peer lookup service. In *LNCS: Peer-to-Peer Systems*, volume 2429/2002, pages 155–165. Springer, Jan 2002.

4. C. Dannewitz, J. Golic, B. Ohlman, and B. Ahlgren. Secure naming for a network of information. In *INFOCOM IEEE Conference on Computer Communications Workshops, 2010*, pages 1–6, march 2010.
5. T. Koponen, M. Chawla, B.-G. Chun, A. Ermolinskiy, K. H. Kim, S. Shenker, and I. Stoica. A data-oriented (and beyond) network architecture. *SIGCOMM Comput. Commun. Rev.*, 37:181–192, August 2007.
6. V. Ramasubramanian and E. G. Sirer. The design and implementation of a next generation name service for the internet. In *Proceedings of the 2004 conference on Applications, technologies, architectures, and protocols for computer communications*, SIGCOMM '04, pages 331–342, New York, NY, USA, 2004. ACM.
7. V. Ramasubramanian and E. G. Sirer. Beehive: O(1)lookup performance for power-law query distributions in peer-to-peer overlays. In *Proceedings of the 1st conference on Symposium on Networked Systems Design and Implementation – Volume 1*, NSDI'04, pages 8–8, Berkeley, CA, USA, 2004. USENIX Association.
8. I. Stoica, D. Adkins, S. Zhuang, S. Shenker, and S. Surana. Internet indirection infrastructure. *SIGCOMM Comput. Commun. Rev.*, 32(4):73–86, 2002.

Chapter 4
Collaborative Web Search

Internet is the largest repository of documents that mankind has ever created. Voluntary contributions from millions of Internet users around the globe, and decentralized, autonomous hosting infrastructure are the sole factors propelling the continuous growth of the Internet. According to the Netcraft (*http://news.netcraft. com/*) Web Server Survey, around 18 million websites were added to the Internet in October 2011 making the total to 504.08 million.

Visibility of a website and its contents is largely governed by the web search engines. However, contemporary Web search services are governed by centrally controlled search engines, which is not healthy for our online freedom due to the following reasons. A Web search service provider can be compromised to evict certain websites from the search results, which can reduce the websites' visibility. Relative ranking of websites in search results can be biased according to the service providers' preference. Moreover, a service provider can record its users' search history for targeted advertisements or spying. For example, the recent PRISM scandal surfaced the secret role of the major service providers in continuously tracking our web search and browsing history.

A decentralized Web search service can subside these problems by distributing the control over a large number of network peers. No single authority will control the search result. It will be computed by combining partial results from multiple peers. Thus a large number of peers have to be compromised to bias a search result. Moreover, a user's queries will be resolved by different peers. All of these peers have to be compromised to accumulate the user's search history.

4.1 Requirements

The usual search semantic in a P2P system is to return all the document names matching the query keywords. In contrast, search results from the contemporary Internet search engines include only a pre-specified number of most relevant results,

R. Ahmed and R. Boutaba, *Collaborative Web Hosting*, SpringerBriefs in Computer Science, DOI 10.1007/978-3-319-03807-0_4, © The Author(s) 2014

along with hyperlinks for fetching additional results on-demand. The requirements for a search mechanism for pWeb can be summarized as follows:

- *Response time*: Centralized search engines can produce results in fraction of a second. We need to ensure a similar response time for pWeb. However, query resolution requires multiple overlay hops in a P2P system. Each overlay hop can require multiple IP hops in the Internet. This can greatly increase the response time for web query resolution.
- *Partial keyword search*: Traditional web search engines offer full text indexing and partial matching of query keywords. However, achieving these two properties in a distributed environment is very challenging and can incur significant network overhead.
- *Distributed ranking*: Web search engines rank the websites in a search result based on their relative importance and relevance to the query keywords. P2P search techniques, on the contrary, return links to all matching contents at once. However, the number of matching websites in a P2P web search result may be very large. Transmitting all of these results to the querying peer will incur significant network and storage overheads. Hence, we need to rank the search results in a distributed manner. This will reduce network and storage overheads by transmitting only the most important and relevant results.
- *Distributed incremental retrieval*: Gradual retrieval of search results in chunks of 10 or 20 records is supported by the contemporary web search engines for performance reasons. For achieving incremental retrieval in a P2P network, the routing mechanism must be able to track previously returned results, and already queried peers. This is well beyond the capabilities of the traditional file-sharing P2P systems.

4.2 Web Search in P2P Networks

None of the previous research works in P2P networks have provided a search mechanism that can fulfill all of the aforementioned requirements. In this section we present the related research works that aim to solve either the similar keyword search problem or the distributed ranking problem in P2P networks.

4.2.1 Similar Keyword Search

4.2.1.1 Structured Techniques

Inability to support partial keyword matching is considered a handicap for DHT-techniques. In the last few years a number of research efforts have focused on extending DHT-techniques for supporting keyword search. Most of these approaches adopted either of the following two strategies:

- Build an additional layer on top of an existing routing mechanism, like Chord [26], CAN [20] or Tapestry [35]. The aim is to reduce the number of DHT lookups per search by mapping related keywords to nearby peers on the overlay. This strategy is proposed in a number of research works including [10, 14, 25] and [28].
- Combine structured and unstructured approaches in some hierarchical manner to gain the benefits of both paradigms. Few research works focus on this strategy including [5, 9] and [27].

A **generic inverted index** on top of a DHT-based network for solving partial-keyword matching has been proposed in [8]. A keyword can be fragmented into η-grams, and each η-gram can be hashed and stored at the responsible peer. This approach can solve partial keyword matching problem in $O(\omega \log N)$ time, where ω is the number of η-grams in a query and N is the number of peers in the system. However, solving the generic DPM problem with this approach will require $O(2^\lambda \log N)$ time, where λ is the number of 1 bits in a query (or advertisement) pattern.

Keyword fusion, presented in [14], is also an inverted indexing mechanism on top of Chord. It supports keyword search only. A document advertised with keywords $\{k_1, k_2, \ldots, k_t\}$ is routed to peers responsible for keys $h(k_1), h(k_2), \ldots, h(k_t)$, where $h(\cdot)$ is the DHT hash function. To reduce the number of DHT-lookups per search, a system-wide dictionary of common keywords is maintained. A query is routed using the most specific keyword and then filtered using the more common keywords specified in the query. In contrast to DPMS and Plexus, advertisement and replication overhead in this system is proportional to the number of keywords associated with the document. A similar inverted indexing mechanism for web page indexing has been proposed in [34].

Joung et al. [10] proposed a distributed indexing scheme, build on a logical, d-dimensional hypercube vector space over a DHT network (they used Chord for their experiment). In this scheme each advertised object is mapped to a d-bit vector according to its keyword set (similar to Bloom filter construction). They treat d-bit vectors as points in d-dimensional hypercube. No restriction on the mapping of a d-dimensional point to a 1-dimensional key space (required for Chord) has been specified. An advertisement is registered to the peer responsible for the d-bit advertisement vector. A query vector (say Q) is computed in the same manner as the advertisement vector. A query is routed to the peers in the Chord ring that are responsible for a key (say P_i) that is a superset of the query vector Q.

The work by Joung et al. [10] and the inverted indexing method presented in [14] represent the two extremes of advertisement and query traffic trade off. In [10], an advertisement is registered at one peer (responsible for the advertised bit vector) and a query is routed to all possible peers that may contain a matching advertisement. In turn, in [14] an advertisement is registered at all the peers responsible for the advertised keywords and the query is routed to the peer responsible for the most uncommon keyword specified in the query.

pSearch [28] utilizes Information Retrieval (IR) techniques on top of CAN (Content-Addressable Network) for facilitating content based full-text search. Keywords associated with an advertised document (or query) are represented as unit vectors. IR techniques like vector space model (VSM) and latent semantic indexing (LSI) are used to compute a unit vector from the keyword list specified with a document (or query). Similarity between a query and a document (or between two documents) is measured using the cosine (i.e., vector dot product) of the vector representation of the corresponding documents or query. Semantically close documents and queries are expected to be mapped to geometrically close point vectors in the Cartesian space. Now the semantic point vectors from LSI or VSM are treated as geometric points in the Cartesian space of CAN. CAN partitions a d-dimensional, conceptual, Cartesian space into zones and assigns each zone to a peer. However this mapping technique uses the same dimensionality for LSI space and CAN. Thus it needs to have a priori knowledge of the possible keywords (or terms) in the whole system. In reality there can be hundreds of possible keywords, and CAN performance degrades at higher dimensions.

Squid [25] has been designed to support partial prefix matching and range queries on top of DHT-based structured P2P networks. In this system Hilbert Space-filling Curve (HSFC) [23] has been used on top of Chord. HSFC is a special type of locality preserving hash function that can map points from a d-dimensional grid (or space) to a 1-dimensional curve in such a way that the nearby points in d-dimensional space are usually mapped to adjacent values on the 1-dimensional curve. Squid converts keywords to base-26 (for alphabetical characters) numbers. A d-dimensional point is constructed from d keywords specified in the query or advertisement. Then a d-dimensional HSFC is used to map a d-dimensional region (i.e., set of points) specified by the query into a set of curve segments in 1-dimension. Finally, each segment is searched using a DHT-lookup followed by a local flooding. Squid supports partial prefix matching (e.g., queries like compu* or net*) and multi-keyword queries; however, Squid does not have provision for supporting true inexact matching of queries like *net*. Another major problem is that the number of (partial) keywords specified in a query or advertisement is bounded by the dimensionality d of the HSFC in use.

4.2.1.2 Non-structured Techniques

Unstructured systems ([1, 2]) identify objects by keywords. Advertisements and queries are expressed in terms of the keywords associated with the shared objects. Structured systems, on the other hand, identify objects by keys, generated by applying one-way hash function on keywords associated with an object. Key-based query routing is much efficient than keyword-based unstructured query routing. The downside of key-based query routing is the lack of support for partial-matching semantics as discussed in the previous section. Unstructured systems, utilizing blind search methods such as *Flooding* [1] and *Random-walk* [16], can easily be modified to support partial-matching queries. But, due to the lack of proper routing information, the generated query routing traffic would be very high. Besides, there would be no guarantee on search completeness.

Many research activities are aimed at improving the routing performance of unstructured P2P systems. Different routing hints are used in different approaches. In [3], routing is biased by peer capacity; queries are routed to peers of higher capacity with higher probability. In [29] and [33], peers learn from the results of previous routing decisions and bias future query routing based on this knowledge. In [4], peers are organized based on common interest, and restricted flooding is performed in different interest groups. Many research works ([3, 13, 33], etc.) propose storing index information from peers within a radius of 2 or 3 hops on the overlay network. All of these techniques reduce the volume of search traffic to some extent, but none provides guarantee on search completeness.

Bloom filters are used by a few unstructured P2P systems for improving query routing performance. In [13] each peer stores Bloom filters from peers one or two hops away. Three ways of aggregating Bloom-filters are also presented. Experimental results presented in [13] show that logical OR-based aggregation of Bloom filters is not suitable for indexing information from peers more than one hop away. In [21] each peer stores a list of Bloom filters per neighbor. The ith Bloom filter in the list of Bloom filters for neighbor M summarizes the resources that are $i - 1$ hops away from neighbor M. A query is forwarded to the neighbor with a matching Bloom filter at the smallest hop-distance. This approach aims at finding the closest replica of a document with a high probability.

4.2.2   Distributed Relevance Ranking

Integrated solutions for distributed web search is not well investigated in the literature, although there exists implementations like YacY and Faroo. Both of these implementations use gossip based index propagation and distributed crawlers. YacY uses distributed PageRank, while Faroo relies on user feedback for ranking search results. Existing research works on distributed ranking can be classified in two broad categories: link structure based ranking and semantic relevance based ranking.

Link structure analysis is a popular technique for ranking. Google uses the PageRank [18] algorithm to compute page weights that measure its authority-ship. Bender et al. [4] proposed a distributed document scoring and ranking system that focuses on correlation between query keywords that appear in query logs. Sankaralingam et al. proposed a P2P PageRank algorithm in [24], where every peer initializes a PageRank score to its local documents and propagates update messages to adjacent peers. DynaRank [11] works in a similar manner, but only propagates update messages when the magnitude of weight change is greater than a threshold value. In JXP [19], each peer computes initial weights for their local pages using standard PageRank and introduces the notion of "external world", which is a logical node representing the outgoing and incoming hyperlinks from the webpages stored in a peer. Each time a peer meets with another peer, it updates the knowledge about its external world. Wang et al. used two types of ranks for overall ranking: *Local PageRank* is computed in each peer based on the standard Pagerank algorithm, and *ServerRank* is computed as the highest local PageRank or the sum of all the PageRanks of a web server [30]. SiteRank [31] computes the rank at the granulaity level of websites instead of web page level using PageRank. Wu et al. proposed a layered Markov model for distributed ranking where links between websites are in the higher layers and links between the web pages within a particular website or domain are in the lower layers [32].

Another research trend is to use Information Retrieval techniques such as VSM (Vector Space Model), which is widely used in centralized ranking systems. However, computing global weight (inverse document frequency or *idf*) in a distributed system is challenging. A random sampling technique is used in [6] to estimate *idf*. In a DHT-based structured network, each keyword is mapped to a particular peer and that peer can compute the approximate value of *idf* [15]. A Gossip-based algorithm is proposed in [17] to approximate both term frequency (*tf*) and *idf* for unstructured P2P networks.

4.3 A Collaborative Approach

This section focuses on Distributed Engine for Web search (DEWS) – a very different approach to decentralized web indexing, ranking and incremental retrieval. Instead of relying on an overlay of regular Internet users, DEWS builds an overlay between the webservers. DEWS exploit the stability in webserver overlay to heavily cache links (network addresses) that are used as routing shortcuts. Thus DEWS can achieve faster lookup, lower messaging overhead, and higher ranking accuracy in search results. DEWS has two fold contributions. First, is achieves a working solution for distributed web search through a novel combination of link caching, route aggregation, distributed ranking and webserver networking. And second, it proposes a novel approach for distributed, bandwidth efficient incremental retrieval. Incremental retrieval is offered by centralized web search engines, yet not supported by other distributed solutions.

This section is organized into three parts. First, the network architecture for DEWS is presented in Sect. 4.3.1. Then the indexing architecture is presented in Sect. 4.3.2. Finally, the process for resolving a distributed Web query is presented in Sect. 4.3.3.

4.3.1 Network Architecture

DEWS uses the Plexus protocol to build an overlay network between the participating webservers. Plexus ensures efficient information lookup along with scalability to network size. It also provides fault-resilience without incurring much replication overhead. In addition, Plexus offers approximate matching between query keywords and webpage keywords, which is not easily achievable by other DHT techniques. DEWS extends the original Plexus protocol in the two ways: (a) route aggregation (b) incremental retrieval. These extensions are explain in the following.

4.3.1.1   Route Aggregation

The aggregate query rate in DEWS is very high. Hence, the number of query message and the associated communication overhead will be very high if these queries are routed one at a time. Instead, DEWS proposes to aggregate multiple queries in same message. This aggregation approach can significantly reduce the number of network messages.

Plexus has an inherent capability of path aggregation for multicast routing. This capability is extended to aggregate multicast traffic from multiple sources. Route aggregation can be explained by the analogy of an airport. Each airport works as a hub. Transit passengers from different sources gather at an airport and depart on different outgoing flights matching their destinations. Similarly, each Plexus node is used as a routing hub.

Algorithm 5 presents the aggregate routing mechanism in DEWS. The default routing mechanism in Plexus is multicasting, since a few peers have to be checked to allow approximate matching. As a result, each message arriving at a peer contains a number of target codewords. Each peer in DEWS is likely to continuously receive query messages, since Web queries from around the globe will be submitted and processed by the system. Instead of instantly forwarding the incoming messages, each peer accumulates incoming messages in a message queue ($msgQ$) for a very small period of time. Target codeword lists ($m.\mathscr{Y}$) in the incoming messages are combined to a master target list $\mathscr{T}$. Then Plexus routing is applied to select the next hop neighbors and the targets in $\mathscr{T}$ are distributed over the selected neighbors. Since, index advertisement and query messages have small size, many of these messages can be packed in a single message and sent to appropriate neighbors. This approach significantly reduces the number of messages in the network.

4.3.1.2   Incremental Retrieval

Incremental retrieval refers to the process of gradually retrieving search results in parts from a repository or server, as offered by the Web search engines. Though it is a challenging problem to achieve incremental retrieval in a distributed setup, an appropriate solution to the problem can save valuable network bandwidth.

DEWS have exploited the Hamming distance based lookup capability of Plexus to achieve incremental retrieval in a distributed manner. In the Plexus search mechanism, list decoding radius ρ can be varied tocontrol the Hamming distance of the

---

Algorithm 5 AggregateRouting in node X

1: Inputs:
        $msgQ$: $\{< pl, \mathcal{Y} >\}$, where pl is message payload
               and $\mathcal{Y}$ is target list for pl.
2: Internals:
        k: Dimension of the linear code $RM(2,m)$
        $\mathcal{L}$: set of neighbors $(X_1, \ldots, X_{k+1})$ and cached links
3: $\mathcal{T} \leftarrow \bigcup_{m \in msgQ} m.\mathcal{Y}$
       {find suitability of each neighbor/link as next hop}
4: $\mathcal{R} \leftarrow \{ \mathcal{T}_w | w \in \mathcal{L} \wedge \mathcal{T}_w \subseteq \mathcal{T} \wedge$
               $(Y \in \mathcal{T}_w \implies w$ is closer to target Y than $X)\}$
5: **while** $\mathcal{T}$ not empty **do**
6:     $\mathcal{O} \leftarrow \phi$
7:     find s such that $\forall \mathcal{T}_w \in \mathcal{R}, |\mathcal{T}_s| \geq |\mathcal{T}_w|$
8:     **for all** $m \in msgQ$ **do**
9:         **if** $m.\mathcal{Y} \cap \mathcal{T}_s \neq \phi$ **then**
10:             $\mathcal{O} \leftarrow \mathcal{O} \cup \{< m.pl, m.\mathcal{Y} \cap \mathcal{T}_s >\}$
11:             $m.\mathcal{Y} \leftarrow m.\mathcal{Y} - \mathcal{T}_s$
12:         **end if**
13:     **end for**
14:     $\mathcal{R} \leftarrow \mathcal{R} - \{\mathcal{T}_s\}$
15:     $\mathcal{T} \leftarrow \mathcal{T} - \mathcal{T}_s$
16:     send $\mathcal{O}$ to peer X_s
17: **end while**

---

discovered advertisement patterns from a query pattern. The edit distance between query and advertisement keywords is proportional to the Hamming distance between the corresponding query and advertisement patterns. DEWS exploits this feature to discover the documents having lesser similarity to the query keywords by gradually increasing ρ.

DEWS gradually explores the peers near a query pattern in steps. For the initial step, it uses a small ρ close to half of the minimum distance between any pair of codewords. For any query, the closest matching advertised keywords can be found within this radius. By increasing the list decoding radius in subsequent steps, it can find additional codewords, further away from the query pattern. The search is repeated with these additional codewords if the user requires additional results or not enough result is found in the first step. For most of the cases, desired number of results can be found in the first step, which can save a lot of network bandwidth.

Unlike expanding ring search or iterative deepening search in unstructured P2P networks, DEWS can find the target peers (i.e., IP addresses) for the next step from the neighbor list of the queried peers in the current step. Hence, the same set of peers is not queried again and again for consecutive steps.

4.3.2 Indexing Architecture

Metrics used for ranking web search results can be broadly classified into two categories: (a) hyperlink structure of the webpages, and (b) keyword to document relevance. Techniques from Information Retrieval (IR) literature are used for measuring relevance ranks. While link structure analysis algorithms like PageRank [18], HITS [12], etc., are used for computing weights or relative significance of each URL. DEWS exploits both of these measures for ranking search results.

4.3.2.1 Hyperlink Index

About 90% hyperlinks in the Web are intra-domain [31]. Topics and ideas in the webpages of a particular website are almost similar or correlated, and it is not reasonable to utilize the authorship of web documents at the level of single pages. Besides, a website is usually reorganized and managed periodically without significant changes in semantics and outgoing hyperlinks to the rest of the Web. The number of websites in the Web is about 100th of the number of webpages. Considering these facts we perform link structure analysis at the granularity level of websites. For the rest of this paper, we use "URL" to refer to the root URL of a website.

Algorithms for computing webpage weights based on hyperlink structure are iterative and require many iterations to converge. In each iteration webpage weights are updated and the new weights are propagated to adjacent URLs for computation in the next iteration. In order to implement such ranking mechanisms on websites, distributed across an overlay network, the adjacency relationships in hyperlink graph has to be preserved while mapping websites to peers. If hyperlinked websites are mapped to the same peer or adjacent peers then network overhead for computing URL weights will be significantly reduced. Unfortunately, there exists no straight forward hyperlink structure preserving mapping of the Web to an overlay network.

DEWS retains the hyperlink structure as a virtual overlay on top of Plexus overlay. It uses a standard shift-add hash function ($\hbar(\cdot)$) to map a website's base URL, say u_i, to a codeword, say $c_k = \hbar(u_i)$. Then Plexus routing is used to lookup $\beta(u_i)$, which is the peer responsible for indexing codeword c_k (Fig. 4.2). For each outgoing hyperlink say u_{it} of u_i, DEWS finds the responsible peer $\beta(u_{it})$ in a similar manner. During distributed link-structure analysis, $\beta(u_i)$ has to frequently send weight update messages to $\beta(u_{it})$. Hence the network address of peer $\beta(u_{it})$ is cached at peer $\beta(u_i)$, which is called a soft-link. Soft-links mitigate the network overhead generated from repeated lookups during PageRank computation. The process of mapping the hyperlink overly over a Plexus overlay is explained in Fig. 4.1.

The index stored in $\beta(u_i)$ has the form $< u_i, w_i, \{< u_{it}, \beta(u_{it}) >\} >$, where w_i is the PageRank weight of u_i. w_i is computed as $w_i = (1 - \eta) + \eta \sum_{t=1}^{g} \frac{w_{it}}{L(u_{it})}$. Here,

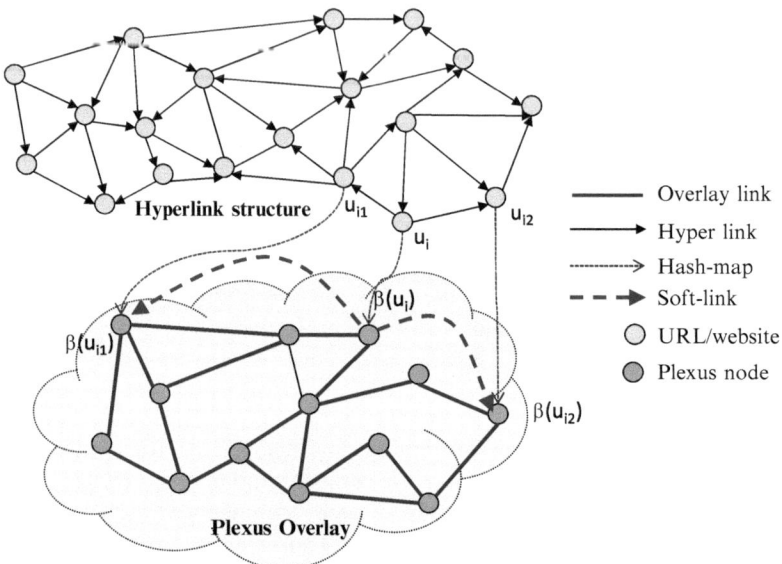

——	Overlay link
——▶	Hyper link
·······▶	Hash-map
− − ▶	Soft-link
○	URL/website
●	Plexus node

Fig. 4.1 Hyperlinks to Plexus overlay mapping

η (usually 0.85) is the damping factor for PageRank algorithm. $\{u_{it}\}$ is the set of webpages linked by u_i and $L(u_{it})$ is the number of outgoing links from webpage u_{it}.

Each peer periodically executes Algorithm 6 to maintain the PageRank weights updated in a distributed manner. To communicate PageRank information between the peers, a PageRank update message is used. This message contains the triplet $< u_s, u_i, \frac{w_s}{L(u_s)} >$, where peer $\beta(u_s)$ sends the message to peer $\beta(u_i)$, and $\frac{w_s}{L(u_s)}$ is the contribution of u_s towards PageRank weight of u_i. Each peer maintains a separate message queue $(\mathcal{Q}_{u_i})$ for each website (u_i) it has indexed. Incoming PageRank messages are queued for a pre-specified period of time and is used to compute the PageRank for each webpage. If the change in newly computed PageRank value is greater than a pre-defined threshold θ, PageRank update messages are sent to $\beta(u_{it})$ for each hyperlinked website u_{it}.

4.3.2.2 Keyword Index

Plexus indexing is used to build an inverted index on the important keywords for each webpage. This index allows us to lookup a query keyword and find all the webpages containing that keyword by forwarding the query message to a small number of peers. Suppose, $\mathcal{K}_i^{rep} = \{k_{ij}^{rep}\}$ is the set of representative keywords for u_i. For each keyword k_{ij}^{rep} in $\mathcal{K}_i^{rep}$, k_{ij}^{dmp} is generated by applying Double Metaphone encoding [7] on k_{ij}^{rep}. Double Metaphone encoding attempts to detect phonetic ('sounds-alike') relationship between words. Motivation behind

---

Algorithm 6 Update PageRank

1: Internals:
  $\mathscr{Q}_{u_i}$: PageRank message queue for u_i
  $L(u_i)$: Number of outlinks for u_i
  w_i: PageRank weight of u_i
  η: Damping factor for PageRank algorithm
  θ: Update propagation threshold
2: **for all** URL u_i indexed in this peer $\beta(u_i)$ **do**
3:   $temp \leftarrow 0$
4:   **for all** $< u_{si}, u_i, \frac{w_{si}}{L(u_{si})} > \in \mathscr{Q}_{u_i}$ **do**
5:     $temp \leftarrow temp + \frac{w_{si}}{L(u_{si})}$
6:   **end for**
7:   $w_i^{new} \leftarrow (1 - \eta) + \eta * temp$
8:   **if** $|w_i^{new} - w_i| > \theta$ **then**
9:     $w_i \leftarrow w_i^{new}$
10:    **for all** out link u_{it} from u_i **do**
11:      send PageRank message $< u_i, u_{it}, \frac{w_i}{L(u_i)} >$ to $\beta(u_{it})$
12:    **end for**
13: **end if**
14: **end for**

---

adapting phonetic encoding is twofold: (i) any two phonetically equal keywords have no edit distance between them, (ii) phonetically inequivalent keywords have less edit distance than the edit distance between the original keywords. In both cases, Hamming distance between the encoded advertisement and search patterns is lesser than that of the patterns generated from the original keywords. This low Hamming distance increases the percentage of common codewords computed during advertisement and search, which eventually increases the possibility of finding relevant webpages.

The process of generating keyword index is depicted in Fig. 4.2. To generate an advertisement or a query pattern P_{ij} from keyword k_{ij}^{rep}, DEWS fragments k_{ij}^{rep} into 3-g ($\{k_{ij}^{rep}\}$) and encode these k-grams along with k_{ij}^{dmp} into a b-bit Bloom filter. This Bloom filter is used as a pattern P_{ij} in $\mathbb{F}_2^b$. Then P_{ij} is *list decoded*[1] it to a set of codewords, $\zeta_\rho(P_{ij}) = \{c_k | c_k \in \mathscr{C} \wedge \delta(P_{ij}, c_k) < \rho\}$, where $\zeta_\rho(\cdot)$ is a list decoding function and ρ is list decoding radius. Finally, Plexus routing is used to lookup and store the index on k_{ij}^{rep} at the peers responsible for codewords in $\zeta_\rho(P_{ij})$. The index for k_{ij}^{rep} is a quadruple $< k_{ij}^{rep}, r_{ij}, u_i, \beta(u_i) >$, where r_{ij} is a measure of semantic relevance of k_{ij}^{rep} to u_i. $\gamma(k_{ij}^{rep})$ represents the set of peers responsible for k_{ij}^{rep}. Evidently, $\gamma(k_{ij}^{rep}) \equiv lookup(\zeta_\rho(BF(\{k_{ij}^{rep}\} \cup \{k_{ij}^{dmp}\})))$, $BF(\cdot)$ represents Bloom filter encoding function.

---

[1] List decoding is the process of finding all the codewords within a given Hamming distance from a (advertisement or query) pattern.

Fig. 4.2 Indexing process
in DEWS

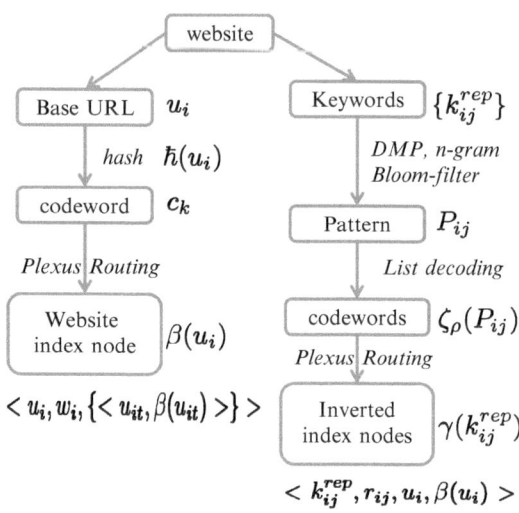

Vector Space Model (VSM) is used for computing the relevance between keyword k_{ij}^{rep} and URL u_i. In VSM, each URL u_i is represented as a vector $v_i = (r_{i1}, \ldots, r_{ig})$, where r_{ij} represents the relevance of the term or keyword k_{ij}^{rep} in u_i, and g is the number of representative keywords in u_i. The relevance weight r_{ij}, of the jth keyword is computed as $tf(k_{ij}^{rep}) * idf(k_{ij}^{rep})$. Here, term frequency $tf(k_{ij}^{rep})$ is the number of occurrences of k_{ij}^{rep} in webpage u_i, while inverse document frequency $idf(k_{ij}^{rep})$ is computed as $idf(k_{ij}^{rep}) = \log \frac{U}{\psi(k_{ij}^{rep})}$. Here, U is the total number of webpages and $\psi(k_{ij}^{rep})$ is the number of webpages containing keyword k_{ij}^{rep}. $tf(k_{ij}^{rep})$ is a measure of the relevance of k_{ij}^{rep} to u_i, while $idf(k_{ij}^{rep})$ is a measure of relative importance of k_{ij}^{rep} w.r.t. other keywords. idf is used to prevent a common term from gaining higher weight and a rare term from having lower weight in a collection.

Computing $tf(k_{ij}^{rep})$ for each keyword $k_{ij}^{rep} \in \mathcal{K}_i^{rep}$ from u_i is straight forward and can be done by analyzing the webpages in u_i. For computing $idf(k_{ij}^{rep})$ one needs to know U and $\psi(k_{ij}^{rep})$. Now, all webpages containing keyword k_{ij}^{rep} are indexed at the same peer. Hence, $\psi(k_{ij}^{rep})$ can be computed by searching the local repository of that peer. However, it is not trivial to compute U in a purely decentralized way. Instead of computing U, the total number of indexed URLs in a peer is used as advocated in [15].

PageRank for URL u_i is computed and maintained in peer $\beta(u_i)$, while the computed PageRank value w_i is used in peers $\gamma(k_{ij}^{rep})$, where a representative keyword k_{ij}^{rep} for webpage u_i is indexed. The Web is continuously evolving and PageRank for the webpages are likely to change over time. As a result, storing PageRank weight, w_i to the peers in $\gamma(k_{ij}^{rep})$ will not be sufficient; it has to be refreshed periodically. To reduce network overhead, softlink to $\beta(u_i)$ is stored in peers $\gamma(k_{ij}^{rep})$. This softlink structure between peers $\beta(u_i)$, $\beta(u_{it})$ and $\gamma(k_{ij}^{rep})$ is presented in Fig. 4.3.

Fig. 4.3 Softlink structure

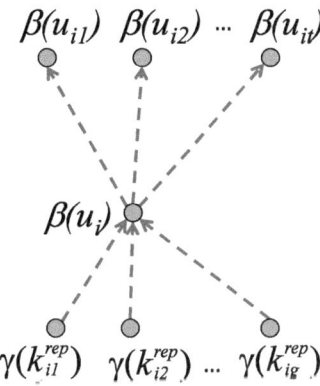

4.3.3 Resolving Web Query

4.3.3.1 Advertising Websites

The pseudocode for advertising a webpage is presented in Algorithm 7. As discussed in the previous two sections, two sets of indexes are maintained for a webpage: (a) using site URL u_i and (b) using representative keywords $\mathcal{K}_i^{rep}$. Lines 3–8 of Algorithm 7 compute the index on u_i, which involves computing the softlinks ($\beta(u_{it})$) for each outgoing hyperlinks from u_i and storing the index in peer $\beta(u_i)$. Lines 9–18 compute the indexes on $\mathcal{K}_i^{rep}$ and advertise the indexes to the responsible peers.

4.3.3.2 Search and Ranking

To resolve Web queries in DEWS, it is fragmented into subqueries – each consisting of a single query keyword, say q_l. Similar to the keyword advertisement process explained in Sect. 4.3.2.2, the Double Metaphone (i.e., q_l^{dmp}) and k-gram ($\{q_l\}$) are computed, and encoded in a Bloom filter P_l. Then the Plexus protocol is used to find the peers responsible for storing the keywords similar to q_l and retrieve a list of triplets like $\{< u_i, w_i, r_{il} >\}$, which gives us the URLs (u_i) containing query keyword q_l along with the link structure weight (w_i) of u_i, and semantic relevance of q_l to u_i, i.e., r_{il}. Now, the querying peer computes the ranks of the extracted URLs using the following equation:

$$rank(u_i) = \sum_{q_l} \sum_{u_i} \vartheta_{il}(\mu \cdot w_i + (1-\mu) \cdot r_{il}) \tag{4.1}$$

In Eq. 4.1, μ is a weight adjustment factor governing the relative importance of link structure weight (w_i) and semantic relevance (r_{il}) in the rank computation process. While ϑ_{il} is a binary variable that assumes a value of one when webpage u_i contains

---

Algorithm 7 Publish webpage

1: Inputs:
       u_i: URL of the webpage to be advertised
2: Functions:
       $\hbar(u_i)$: hash map u_i to a codeword
       $\gamma_r(P)$: $\{c_k | c_k \in \mathscr{C} \wedge \delta(P, c_k) \leq r\}$
       $lookup(c_k)$: Finds the peer that stores c_k
3: $\beta(u_i) \leftarrow lookup(\hbar(u_i))$
4: **for all** out-link u_{it} of $\{u_i\}$ **do**
5:    $\beta(u_{it}) \leftarrow lookup(\hbar(u_{it}))$
6: **end for**
7: $w_i \leftarrow$ initial PageRank of u_i
8: store $< u_i, w_i, \{u_{it}, \beta(u_{it})\} >$ to peer $\beta(u_i)$
9: $\mathscr{K}_i^{rep} \leftarrow$ set of representative keywords of u_i
10: **for all** k_{ij}^{rep} in $\mathscr{K}_i^{rep}$ **do**
11:    $k_{ij}^{dmp} \leftarrow DoubleMetaphoneEncode(k_{ij}^{rep})$
12:    $P_{ij} \leftarrow BloomFilterEncode(\{k_{ij}^{rep}\} \cup \{k_{ij}^{dmp}\})$
13:    $r_{ij} \leftarrow$ relevance of k_{ij}^{rep} to u_i
14:    **for all** c_k in $\zeta_\rho(P_{ij})$ **do**
15:      $v \leftarrow lookup(c_k)$
16:      store $< k_{ij}^{rep}, r_{ij}, u_i, \beta(u_i) >$ to peer v
17:    **end for**
18: **end for**

---

keyword q_l and zero otherwise. While the implication of simply summing w_i and r_{il} together is not obvious, similar approaches were proposed in [22]. Although, one can devise complicated ways to combine these two measures together, a simple summation suffices to achieve the desired effect.

The query process in DEWS is explained in Algorithm 8. In this algorithm, a separate $lookup(c_k)$ is required for each target codeword c_k. In practice separate lookup of each target is expensive in terms of network usage. Instead, DEWS uses the extended multicast routing mechanism with route aggregation as explained in Sect. 4.3.1.1.

4.4   Summary

DEWS is a self-indexing architecture for the Web. It enables webservers to collaboratively index the Web and respond to Web queries in a completely decentralized manner. In DEWS, we have the provision for approximate matching on query keywords, and distributed ranking on semantic relevance and link-structure characteristics. Network and storage overheads for achieving this decentralization is not significant. DEWS scales well with network size and the number of indexed webpages. In addition, the ranking accuracy of DEWS is comparable to the ranking accuracy of the centralized ranking solution. The route aggregation technique,

---

Algorithm 8 Query

Input:
    Q: set of query keywords $\{q_l\}$
    T: Most relevant T webpages requested
Internals:
    μ: Weight adjustment on link-structure vs relevance
    ρ: list decoding radius
$\xi \leftarrow$ *empty associative array*
for all $q_l \in Q$ **do**
    $q_l^{dmp} \leftarrow DoubleMetaphoneEncode(q_l)$
    $P_l \leftarrow BloomFilterEncode(\{q_l\} \cup \{q_l^{dmp}\})$
    for all $c_k \in listDecode_\rho(P_l)$ **do**
        $n \leftarrow lookup(c_k)$
        for all $\{< u_i, w_i, r_{il} >\} \in n.retrive(q_l)$ **do**
            $\xi[u_i].value \leftarrow \xi[u_i].value + \mu \cdot w_i + (1 - \mu) \cdot r_{il}$
        end for
    end for
end for
sort ξ based on *value*
return top T u_i from ξ

---

proposed in DEWS, outperforms the original Plexus routing protocol in terms of network usage efficiency. DEWS is highly resilient to peer failures due to the existence or alternate routing paths and smart replication policy. Neither routing efficiency nor ranking accuracy degrades significantly even in presence of 20% failures. Compared to a centralized solution, DEWS will incur some network overhead. In exchange DEWS will give us the freedom of searching and exploring the Web without any control or restrictions, as can be imposed by the contemporary centrally controlled search engines.

References

1. Gnutella website, http://www.gnutella.com.
2. The FastTrack Peer-to-Peer technology, http://www.fasttrack.nu/.
3. Y. Chawathe, S. Ratnasamy, L. Breslau, N. Lanham, and S. Shenker. Making Gnutella-like P2P systems scalable. pages 407–418, 2003.
4. E. Cohen, A. Fiat, and H. Kaplan. Associative search in Peer-to-Peer networks: Harnessing latent semantics. 2003.
5. P. Ganesan, Q. Sun, and H. Garcia-Molina. Adlib: A self-tuning index for dynamic Peer-to-Peer systems. In *Proc. of the International Conference on Data Engineering (ICDE)*, pages 256–257, Los Alamitos, CA, USA, 2005. IEEE Computer Society.
6. V. Gopalakrishnan, R. Morselli, B. Bhattacharjee, P. Keleher, and A. Srinivasan. Distributed ranked search. In *HiPC*, pages 7–20, 2007.
7. M. R. Haque, R. Ahmed, and R. Boutaba. Qpm: Phonetic aware p2p searching. In *IEEE Peer-to-Peer Computing*, pages 131–134, 2009.

8. M. Harren, J. M. Hellerstein, R. Huebsch, B. T. Loo, S. Shenker, and I. Stoica. Complex queries in DHT-based Peer-to-Peer networks. In *Proc. of International Workshop on Peer-to-Peer Systems (IPTPS)*, pages 242–259, 2002.

9. X. Jin, W.-P. K. Yiu, and S.-H. Chan. Supporting multiple-keyword search in a hybrid structured Peer-to-Peer network. In *Proc. of IEEE International Conference on Communications (ICC)*, pages 42–47, Istanbul, June 2006.

10. Y. Joung, L. Yang, and C. Fang. Keyword search in DHT-based Peer-to-Peer networks. *IEEE Journal on Selected Areas in Communications (JSAC)*, 25(1):46–61, January 2007.

11. M. Kale and P. S. Thilagam. DYNA-RANK: Efficient Calculation and Updation of PageRank. In *ICCSIT*, pages 808–812, 2008.

12. J. Kleinberg. Authoritative sources in a hyperlinked environment. *Journal of the ACM (JACM)*, 46(5):604–632, 1999.

13. M. Li, W. Lee, and A. Sivasubramaniam. Neighborhood signatures for searching P2P networks. In *Proc. of Seventh International Database Engineering and Applications Symposium (IDEAS)*, pages 149–159, 2003.

14. L. Liu, K. D. Ryu, and K. Lee. Supporting efficient keyword-based file search in Peer-to-Peer file sharing systems. In *Proc. of GLOBECOM*, 2004.

15. Z. Lu, B. Ling, W. Qian, W. S. Ng, and A. Zhou. A distributed ranking strategy in P2P based IR systems. In *APWeb*, pages 279–284, 2004.

16. Q. Lv, P. Cao, E. Cohen, K. Li, and S. Shenker. Search and replication in unstructured Peer-to-Peer networks. In *Proc. of the International Conference on Supercomputing (ICS)*, pages 84–95, 2002.

17. R. Neumayer, C. Doulkeridis, and K. Nørvåg. A hybrid approach for estimating document frequencies in unstructured p2p networks. *Inf. Syst.*, 36(3):579–595, May 2011.

18. L. Page, S. Brin, R. Motwani, and T. Winograd. The pagerank citation ranking: Bringing order to the web. Technical report, Stanford Digital Library Technologies Project, 1999.

19. J. X. Parreira and G. Weikum. JXP: Global Authority Scores in a P2P Network. In *8th Int. Workshop on Web and Databases (WebDB)*, 2005.

20. S. Ratnasamy, P. Francis, M. Handley, R. Karp, and S. Shenker. A scalable content-addressable network. *SIGCOMM Comput. Commun. Rev.*, 31:161–172, August 2001.

21. S. Rhea and J. Kubiatowicz. Probabilistic location and routing. In *Proc. of IEEE INFOCOM*, 2002.

22. M. Richardson and P. Domingos. The Intelligent surfer: Probabilistic Combination of Link and Content Information in PageRank. In *NIPS*, pages 1441–1448, 2001.

23. H. Sagan. *Space-filling curves*. Springer-Verlag, 1994.

24. K. Sankaralingam, M. Yalamanchi, S. Sethumadhavan, and J. Browne. Pagerank computation and keyword search on distributed systems and p2p networks. *Journal of Grid Comp.*, 1(3):291–307, 2003.

25. C. Schmidt and M. Parashar. Enabling flexible queries with guarantees in P2P systems. *IEEE Internet Computing*, 8(3):19–26, June 2004.

26. I. Stoica, R. Morris, D. Liben-Nowell, D. R. Karger, M. F. Kaashoek, F. Dabek, and H. Balakrishnan. Chord: a scalable Peer-to-Peer lookup protocol for Internet applications. *IEEE/ACM TON*, 11(1):17–32, 2003.

27. C. Tang and S. Dwarkadas. Hybrid global-local indexing for efficient Peer-to-Peer information retrieval. In *Proc. of the Symposium on Networked Systems Design and Implementation (NSDI)*, June 2004.

28. C. Tang, Z. Xu, and M. Mahalingam. pSearch: information retrieval in structured overlays. *ACM SIGCOMM Computer Communication Review*, 33(1):89–94, 2003.

29. D. Tsoumakos and N. Roussopoulos. Adaptive probabilistic search for Peer-to-Peer networks. 2003.

30. Y. Wang and D. DeWitt. Computing pagerank in a distributed internet search system. In *Proc. VLDB*, volume 30, pages 420–431, 2004.

31. J. Wu and K. Aberer. Using siterank for decentralized computation of web document ranking. In *AH*, pages 265–274, 2004.

32. J. Wu and K. Aberer. Using a Layered Markov Model for Distributed Web Ranking Computation. In *Proc. ICDCS*, pages 533–542, Jun. 2005.
33. B. Yang and H. Garcia-Molina. Improving search in Peer-to-Peer networks. 2002.
34. K.-H. Yang and J.-M. Ho. Proof: A DHT-based Peer-to-Peer search engine. In *IEEE/WIC/ACM International Conference on Web Intelligence (WI 2006)*, pages 702–708, December 2006.
35. B. Zhao, L. Huang, J. Stribling, S. Rhea, A. Joseph, and J. Kubiatowicz. Tapestry: a resilient global-scale overlay for service deployment. *IEEE JSAC*, 22(1):41–53, Jan. 2004.

Chapter 5
Availability

Since its inception, peer-to-peer (P2P) technology has been applied for numerous distributed applications, including file sharing, distributed computing, multi-player gaming, media streaming and instant messaging. None of these applications require or assume a persistent service guarantee from the underlying P2P overlay. Yet there exists other applications like web hosting, online backup, content distribution etc., that require persistence in resource/service availability. P2P systems rely on commodity machines, voluntarily participating at the network edge. As a result, it is challenging to use P2P technology for deploying any application that requires persistent resource/service availability.

Existing proposals in P2P systems use replication as the primary means for increasing resource availability. Replication strategies in P2P systems can be broadly classified as time-based replication and quantitative replication. In quantitative replication approaches availability is ensured by consistently maintaining a fixed number of replicas per resource. On the other hand, time-based replication approaches utilize a peer's uptime history to reuse a replica from the peer's previous session.

In quantitative replication approaches, content availability is proportional to the number of its replicas. But, increasing the number of replicas has a number of side effects. First, it incurs increased network overhead for replica placement and update propagation between the replicas when the original content is updated. Second, storage overhead increases linearly with the number of replicas. Third, it requires additional mechanisms for keeping track of the replicas for efficient query forwarding. And last but not the least, query load balancing among the replicas of a specific content becomes an important issue from fairness point of view. Existing availability approaches [1,2,16,19] that solely rely on replication are either bandwidth hungry or require complex predictive knowledge for replica updates and relocation. These approaches frequently burden the peers with longer uptime, which results into a skewed load distribution and a negative impact on availability.

Time-based replication strategies, on the other hand, utilize daily uptime behavior of the peers to replicate a content. Cyclic diurnal pattern in peer availability has been observed in a number of previous studies including [6, 8, 15, 20].

R. Ahmed and R. Boutaba, *Collaborative Web Hosting*, SpringerBriefs in Computer Science, DOI 10.1007/978-3-319-03807-0_5, © The Author(s) 2014

Rzadca et al. [13] have shown that diurnal behavior of peers can be a useful characteristic for improving availability if the system has a truly global scope. For example, consider two peers separated by 12-h difference in time zone. They will exhibit complementary availability patterns, if both of them remain online during daytime and off-line at night. Even for the peers located in the same time zone, mutually exclusive availability patterns may be observed due to their Internet usage habits or job nature.

In this chapter, we investigate the availability requirements for P2P web hosting in Sect. 5.1. Then we present the existing solutions for improving availability in P2P networks (Sect. 5.2). In Sects. 5.3 and 5.4, we present a globally optimized and efficient protocol, named *S-DATA (Structured approach for Diurnal Availability by Temporal Assemblage)* that maximizes 24/7 content availability in a P2P network. S-DATA minimizes the aforementioned shortcomings of the existing time-based availability schemes.

5.1 Requirements

Here outline the requirements on the group formation process and replication mechanism. A good solution for ensure availability for pWeb should fulfill these requirements.

- *Replication strategy*: As discussed before, both quantitative and time-based replication strategies have their relative advantages and disadvantages. Time-based replication is more appropriate for achieving smaller group size and for reducing network overhead. However, for ensuring content availability at any given time, we have to perform quantitative replication as well. Hence, a hybrid replication strategy is required pWeb.
- *Group size*: In a P2P network of a million peers, it is a challenging problem to match and tie peers in small groups in such a way that the maximum availability can be achieved with minimal replication overhead. Group size should be as small as possible in order to reduce the replication overhead.
- *Global optimization*: the group formation process should be globally optimized and should not incur significant network overhead. Existing time-based availability approaches ([3, 13, 17]) rely on unstructured, gossip-based protocol and do not deliver guarantee on the above mentioned requirements.
- *Continuity*: At any given time, at least one peer (or a pre-specified number of peers) should be online within a group. This would ensure a content's availability regardless of its popularity.

5.2 Availability in P2P Systems

A number of approaches to improve availability in P2P systems can be found in the literature. To the best of our knowledge, only a few of these approaches focus on increasing content availability by adopting a time-based replication strategy. In [17], we proposed the *DATA* protocol that construct replication groups using complementary availability patterns of peers through a gossip based routing technique applicable to unstructured networks. Blond et al. [3] proposed two availability-aware applications that take into account the peers' previous availability history collected through an epidemic protocol. Using a simple predictor, they propose to find the best matching peer to meet the specific goals of the application.

A group based Chord model is proposed in [7] to minimize the impact of frequent arrivals and departures of peers. The redundancy group based approach proposed by Schwarz et al. [16] tries to improve availability by utilizing the cyclic behavior of peers distributed across the World. They proposed a hill-climbing strategy to determine redundancy groups for data objects. They proposed a counter-based availability score update mechanism through periodic scans. However, the counter mechanism cannot consistently capture phase relationships within a peer and between peers. For example, a peer having diurnal availability pattern will be online for the longest consecutive stretch starting in the morning, when its counter is the lowest. But this fact is not reflected in their mechanism.

Rzadca et al.[13] proposed to represent peer availability as a function of discrete time to minimize the number of replicas. In their model, availability is represented by a set of time slots in which a peer is available with certainty, i.e., the used discrete on-off availability. In contrast, S-DATA represents availability by historical probability at discrete time slots. Our probabilistic model captures diurnal availability patterns more accurately, since peer connectivity cannot be predicted with absolute certainty in a real world network. Moreover, the group formation approach proposed in [13] uses a single-valued scoring function, which only considers the number of newly covered slots while making group formation decision. On the contrary, our utility function considers relative improvement from both sides and considers the size of the resulting group. Finally, their model only targets to ensure 1-availability across time slots, whereas *S-DATA* proposes to consider β-Availability to provide better reliability.

A significant number of research works aim to increase availability by adopting various strategies for quantitative replication. These works vary in the type of redundancy, method of replica regeneration, and the number and location of peers storing redundant data. Bhagwan et al. explored the issues of replication granularity, replica placement, and application characteristics in [1]. In terms of replica generation approach, redundancy is achieved in two ways: (i) replicating the complete data, and (ii) fragmenting and encoding the data by network coding in such a way that not all fragments are needed to reproduce the original content [1, 16]. Data replication is mainly done in two ways: reactive [2] or proactive [19]. Both of these approaches aim to optimally place the replicas in a minimum number of peers so that the overall content availability remains high. Existing solutions use information like peers' session time and churn [10], availability history [3], lifespan distribution [5], machines' uptime, downtime, lifetime, and correlation among them [4], Mean Time to Failure [9], up time score [16], recent up time [14], application specific availability [18], availability-prediction guided replica placement [2, 11], and probabilistic models [12] to reduce redundancy while retaining high availability. These approaches rely on quantitative replication, whereas S-DATA combines both time-based and quantitative replication strategies. Another major difference of S-DATA with these schemes is that they make no distinction between transient and permanent disconnections and data stored at a peer is reused upon its return to the system. Ignoring stored data after peers' offline period incurs significant network overhead, which S-DATA can readily avoid by co-relating a returning peer with its previous session.

5.3 Conceptual Overview

5.3.1 Architecture

As depicted in Fig. 5.1, S-DATA architecture evolves around three conceptual components: *replication group*, *Group Index Overlay* (GIO) and *Content Index Overlay* (CIO). Replication groups provide a persistent storage by exploiting diurnal uptime-behavior of the regular peers. GIO maintains peers' and groups' availability information. While CIO retains an indirect mapping from content name to content location. In the following we explain each of these three components.

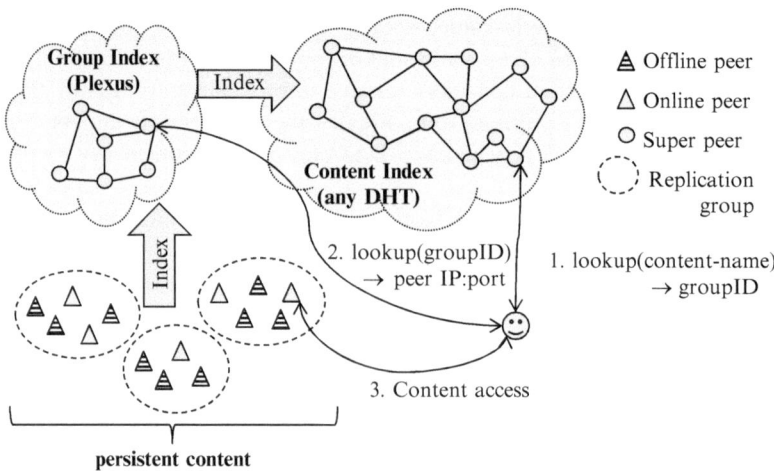

Fig. 5.1 Conceptual architecture of S-DATA

Replication group: In S-DATA, peers are clustered into small groups based on their diurnal availability pattern. Within a replication group, peers have mutually exclusive uptime with little overlap. In a replication group with β-availability, it is ensured that at least β members from that group will be online at any given time. Each member within a group replicates each others contents, and works as a proxy for the off-line members of that group.

Group index overlay: It has two functions. First, during group formation, it works as a distributed agent for match-making peers with complementary uptime behavior. Second, it acts as an indirection structure during content lookup. Initially each peer advertises its availability pattern as a bit-vector to this overlay. During group formation, peers willing to form a group search for other peers (or groups) having complementary uptime behavior. To the best of our knowledge, Plexus is the only Distributed Hash Table (DHT) technique that supports approximate bit-vector matching in an efficient manner. Hence, we used Plexus as the indexing and routing protocol for GIO. At any given time, this overlay maps a group ID to one (or β) online peer from that group.

Content index overlay: This overlay can be implemented using any DHT-technique depending on the application-specific requirements. This overlay maps a content name to a group ID. In order to search and download a content, a peer will first search the CIO and discover a group ID. Then it will lookup the group ID in the GIO and find the location (IP:port) of an alive peer currently hosting that content and download it. Mapping a content name to a group ID, instead of directly mapping to a peer ID incurs an additional lookup. But, this lookup is necessary to facilitate dynamically associate a content name to the currently online peer hosting that content.

From uptime point of view, we assume that the peers in a replication group are regular peers with moderate online time (4–8 h) on a daily basis. While the peers in the indexing overlays are superpeers with longer uptime, higher communication bandwidth and storage capacity.

5.3.2 Availability Vector

The traditional definition of peer availability is simply measured by the fraction of time a peer is online [2] within a certain time period. If a peer joins and leaves m times during a period of T hours, and every time remains up for t_k hours, then its availability can be computed as, $\frac{\sum_{k=1}^{m} t_k}{T}$. This formula does not take the diurnal availability pattern in peer uptime behavior into account. This fact has been mathematically proven by Yang et al. in [21].

In S-DATA, we divide 24-h of a day in K equal-length time-slots w.r.t. GMT+0, and estimate the probability of a peer being online in each time-slot based on its historical behavior. Thus the availability of a peer, say x, is defined as $\mathscr{A}_x = \{a_{x1}, a_{x2}, \ldots, a_{xk}, \ldots, a_{xK}\}$, where $\mathscr{A}_x$ is the K-dimensional availability vector for peer x, and a_{xk} is the probability of peer x being online in slot k.

The responsibility of computing and maintaining the availability vectors can be dedicated either to the P2P client software or to GIO. Each of these alternatives has its own merits and demerits, and can be considered as an implementation specific choice. Computing and maintaining availability vectors at the client software will give a more accurate estimate of a peer's availability vector, and will generate minimal network traffic. However, a client software can be maliciously modified to report a fake availability vector. Alternatively, the availability vectors can be computed and maintained at GIO. This approach can generate more reliable probability values for the availability vectors, though at the expense of increase network traffic and decreased accuracy of the computed availability.

5.4 S-DATA Protocol Details

5.4.1 Terminology

In S-DATA we use four indexes (see Table 5.1) for group formation and content lookup. $\mathscr{I}_e$ represents an indexing peer in GIO, which is responsible for storing the ID of e (ID_e), where e can be a regular peer or a group. $\mathscr{I}_e$ works as e's proxy for meta-information exchange. For a regular peer, say x, $\mathscr{I}_x$ stores an $\mathscr{M}_x$ record, which contains the availability vector (A_x), ID (ID_x) and network location (Loc_x) for x, as well as the group ID (ID_{G_x}) and index location ($\mathscr{I}_{G_x}$) for x's group G_x. For a group G, $\mathscr{I}_G$ contains index record $\mathscr{N}_G$, which contains group availability vector ($\mathscr{A}_G$), group ID (ID_G), and for each member x of G, its ID (ID_x), index location ($\mathscr{I}_x$) and network location (Loc_x). To enable approximate matching between peers' and groups' availability vectors, we maintain $\mathscr{V}_e$ indexes that contain availability pattern (S_e, explained in Sect. 5.4.2.1), availability vector ($\mathscr{A}_e$), ID (ID_e) and index location ($\mathscr{I}_e$) for e. $\mathscr{V}_e$ is stored in all peers $\mathscr{L}_e$ within a pre-specified Hamming distance from S_e. Finally, for content lookup another set of indexes ($\mathscr{K}_w$) is maintained in CIO. For each keyword w attached to a content an index ($\mathscr{K}_w$) is stored in CIO at peer $\mathscr{J}_w$, which is responsible for keyword w. $\mathscr{K}_w$ retains the content's ID (ID_{doc}), other keywords describing the content ($\{w_i\}$), group ID (ID_G) and index location ($\mathscr{I}_G$) of the group that hosts the content.

Table 5.1 List of indexes in S-DATA

Name	Overlay	Indexed information
$\mathscr{M}_x$	GIO/$\mathscr{I}_x$	$< \mathscr{A}_x, ID_x, Loc_x, ID_{G_x}, \mathscr{I}_{G_x} >$
$\mathscr{N}_G$	GIO/$\mathscr{I}_G$	$< \mathscr{A}_G, ID_G, \{< ID_x, \mathscr{I}_x, Loc_x > | x \in G\} >$
$\mathscr{V}_e$	GIO/$\mathscr{L}_{S_e}$	$< S_e, \mathscr{A}_e, ID_e, \mathscr{I}_e >$
$\mathscr{K}_w$	CIO/$\mathscr{J}_w$	$< ID_G, \mathscr{I}_G, ID_{doc}, \{w_i | w_i \in doc\} > \mathscr{L}_{\hat{S}_x}$

5.4.2   Indexing Availability Information

To cluster regular peers in globally optimized replication groups, we need to index each peer's availability information ($\mathcal{V}_e$) to GIO. This indexing process involves two steps: (i) encoding availability vector ($\mathcal{A}_e$) to bit-vector (S_e) and (ii) advertisement using Plexus protocol. These two steps are explained in the following.

5.4.2.1   Availability Vector Encoding

It can be easily seen that the availability vector $\mathcal{A}_i$ is a K-dimensional vector of uptime probabilities, whereas the advertisement (or query) patterns in a Plexus network built on an $< n, k, d >$ code are n-bit values. Hence, we need a means to encode a K-dimensional availability vector into an n-bit pattern.

In S-DATA we have used $K = 24$ slots for availability vector. While for Plexus implementation, we have used the $< 24, 12, 8 >$ Extended Golay Code G_{24}. Trivially, we can directly encode each probability value a_{ik} in $\mathcal{A}_i$ to 1-bit in the 24-bit advertisement (or query) pattern. We can use a threshold, say θ, and can set the k-th bit of the 24-bit encoded pattern to 1 if $a_{ik} > \theta$. Unfortunately, this encoding will incur significant information loss and will degrade the approximate matching performance in Plexus network.

Alternatively, we use a better encoding scheme based on the observation that consecutive values in the availability vector are usually similar in magnitude. To exploit this observation, we average the probability values in two adjacent slots, and obtain a 12-dimensional availability vector $\acute{\mathcal{A}}_i = \{\acute{a}_{i1}, \acute{a}_{i2}, \ldots \acute{a}_{i12}\}$, where $\acute{a}_{ij}$ is computed as $\acute{a}_{ij} = \frac{(a_{i(2j-1)} + a_{i(2j)})}{2}$. Now, we encode each $\acute{a}_{ij}$ into two bits in the 24-bit advertisement pattern as follows. $\acute{a}_{ij}$ is encoded to 00 if $\acute{a}_{ij}$ is less than $\frac{1}{3}$. If $\acute{a}_{ij}$ is between $\frac{1}{3}$ and $\frac{2}{3}$ then the encoding is 01. Otherwise, $\acute{a}_{ij}$ is greater than $\frac{2}{3}$ and is encoded to 11. This encoding reflects the numeric distance in $\acute{a}_{ij}$ to the Hamming distance in advertisement patterns. The bit-vector obtained from this encoding is referred to as S_e for element e. S_e is required to advertise and search peer availability information in the GIO.

5.4.2.2   Advertisement

An advertising peer, say x, first computes the n-bit advertisement pattern, say S_x, as explained above. Then x sends the tuple $< \mathcal{S}_x, \mathcal{A}_x, ID_x, Loc_x, ID_{G_x}, \mathcal{I}_{G_x} >$, to $\mathcal{I}_x$. In this advertisement ID_{G_x} and $\mathcal{I}_{G_x}$ will be empty, if x has not formed a group. Upon receiving the advertisement message, $\mathcal{I}_x$ computes the codewords within a pre-specified Hamming distance from S_x and uses Plexus routing to route and index the advertisement ($\mathcal{V}_x$) to the peers ($\mathcal{L}_{S_x}$) responsible for these codewords.

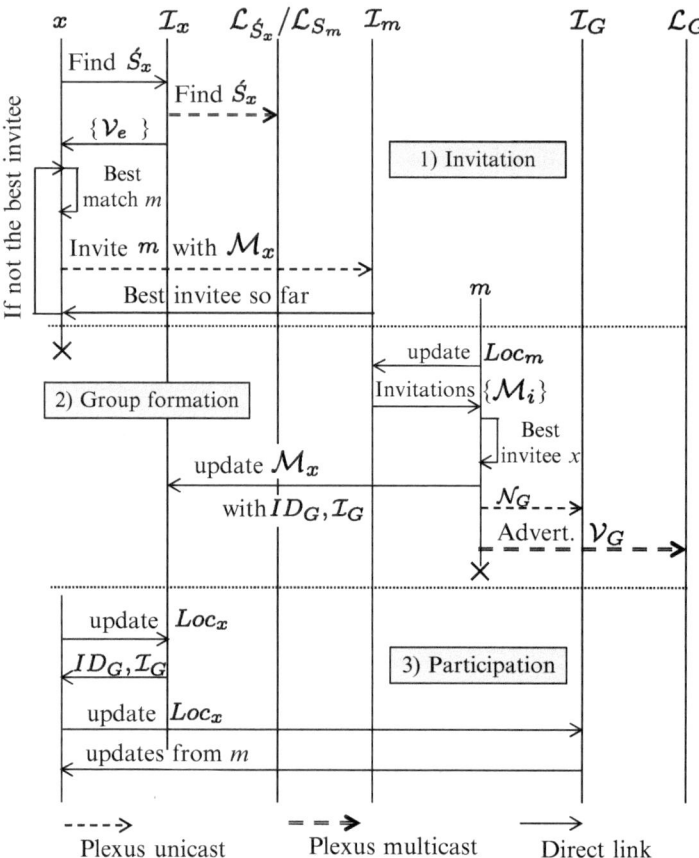

Fig. 5.2 Sequence diagram show group formation of x with m

5.4.3 Group Formation

This process lies at the core of S-DATA protocol. Our target is to cluster peers into groups in such a way that the group sizes are minimal and at any given time at least $\beta \geq 1$ peers from a group is online with the highest possible probability.

The most challenging part of this process is to relay group formation messages between peers that may not be simultaneously online. To this end, we use GIO as a message relay. Figure 5.2 presents a sequence of message exchanges between indexing peers in GIO and regular peers x and m while forming a 1-availability group G. It is worth noting that x and m are not online simultaneously and hence they have no direct message exchange. The Group formation process is composed of the following three steps:

1. *Invitation*: We assume that on average a regular peer will be online for L time-slots on a daily basis. It will be the responsibility of a peer to maintain β peers in its group during the L-slots it is online and the next L-slots. To find a suitable peer that can improve its own group's availability for the next L-slots, peer x computes an availability pattern $\acute{S}_x$. $\acute{S}_x$ has bits $t + L + 1$ to $t + 2L$ set to 1, assuming that the availability pattern S_x of peer x has bits t to $t + L$ set to 1. Once $\acute{S}_x$ is computed, peer x forwards it to $\mathscr{I}_x$. $\mathscr{I}_x$ uses Plexus multi-cast routing to find the peers ($\mathscr{L}_{\acute{S}_x}$) in GIO responsible for indexing peer/group availability records ($\mathscr{V}_e$) similar to $\acute{S}_x$. From the availability records ($\mathscr{V}_e$) returned by $\mathscr{I}_x$, peer x selects the most appropriate peer, say m, that maximizes its group's availability. Peer x locates the indexing peer ($\mathscr{I}_m$) for m using Plexus routing and sends an invitation request to $\mathscr{I}_m$ that includes the $\mathscr{V}_x$ record.
2. *Group formation*: Upon becoming online m updates $\mathscr{I}_m$ with its new network location (Loc_m). In response $\mathscr{I}_m$ sends all the invitations ($\{\mathscr{V}_e\}$) for m that has been accumulated during m's offline period. Among these invitations, m selects the best candidate x. If x is already a member of an existing group then m simply joins the group otherwise it creates a new group G. To create or update the group index in GIO, m may require to transmit three messages: (a) if m created a new group, then it has to update the $\mathscr{M}_x$ record in $\mathscr{I}_x$ so that x can learn about G upon returning; (b) m has to index ($\mathscr{V}_G$) to all peers ($\mathscr{L}_G$) within a certain Hamming distance from S_G; (c) finally, m has to store the group index $\mathscr{N}_G$ to $\mathscr{I}_G$.
3. *Participation*: During its next online session peer x will update I_x with its new network location Loc_x. If the previous invitation from x was honored by m then I_x responds with the newly formed group's information (ID_G and $\mathscr{I}_G$). x updates $\mathscr{I}_G$ with its location information Loc_x. $\mathscr{I}_G$ responds with any update from m or other members of G. On the other hand, if the invitation from x was not accepted by m, then x has to restart the group formation process with the next best matching peer, other than m.

The above mentioned process of forming 1-availability group can be easily extended to construct β-availability groups. Two modifications in Step 1 of the above process are required. First, x should be the highest ID peer among the online members of its group (G_x). And second, x should send invitations to $\beta - f$ peers simultaneously, where f is the number of peers in x's group who shall be online in the L-time slots following the online period of x.

5.4.4 Group Maintenance

The diurnal availability pattern of a peer may change over time. In such a situation, a peer, say x, may want to change its group. Group changing involves leaving the current group and joining a new group. The process of joining a group has been described in Sect. 5.4.3. To leave its current group G_x, peer x has to update two peers in GIO. First, x has to remove its index information from $\mathscr{N}_{G_x}$ record, which is stored in peer $\mathscr{I}_{G_x}$. And second, x has to clear the ID_{G_x} and $\mathscr{I}_{G_x}$ fields in $\mathscr{M}_x$ record, which is stored in $\mathscr{I}_x$. It should be noted that we use soft-state registration for advertising $\mathscr{V}_x$ records to $\mathscr{L}_{S_x}$. Hence, the $\mathscr{V}_x$ records will be automatically removed from the peers in $\mathscr{L}_{S_x}$, if x does not re-advertise before the previous advertisement expires.

5.4.5   Content Indexing and Lookup

In the following we describe the mechanisms for content indexing and lookup.

5.4.5.1   Content Indexing

Traditionally a content in a P2P network is tagged with a set of descriptive keywords, ($w \in \{w_i\}$). These keywords are used to locate the peers(s) ($\mathscr{J}_w$) in CIO for storing the $\mathscr{K}_w$ record. While advertising a content a peer, say x, may or may not be a member of a replication group. If x is a member of a replication group, say G_x then ID_{G_x} and $\mathscr{I}_{G_x}$ are stored in $\mathscr{K}_w$ record, otherwise ID_x and $\mathscr{I}_x$ are used. However, $\mathscr{K}_w$ is not updated when x forms a group. Rather, $\mathscr{K}_w$ is updated in a reactive manner during content lookup. This process is described in the following section.

5.4.5.2   Content Lookup

A query for keyword w will be routed to $\mathscr{J}_w$ using the routing protocol in CIO. Based on the information found in $\mathscr{K}_w$, the query will be forwarded to either $\mathscr{I}_{G_x}$ if the content host x has formed a group and $\mathscr{K}_w$ has been updated, or the query will be forwarded to $\mathscr{I}_x$. In a regular scenario, the query will be forwarded to $\mathscr{I}_{G_x}$ and the location Loc_y of the currently alive peer y in G_x will be return to the querying peer via $\mathscr{J}_w$. On the other hand, if x has formed a group but $\mathscr{K}_w$ has not been updated, then $\mathscr{J}_w$ will contact $\mathscr{I}_x$, which will respond with ID_{G_x} and $\mathscr{I}_{G_x}$. Accordingly, $\mathscr{J}_w$ will update $\mathscr{K}_w$ for future references. Finally, $\mathscr{J}_w$ will contact $\mathscr{I}_{G_x}$ to obtain the location (Loc_y) of the currently active peer (y) in G_x.

5.5   Summary

In this chapter, we have identified the availability requirement for P2P web hosting. A short survey of the existing approaches for ensuring availability in P2P file-sharing applications has been presented as well. Finally, we have presented an efficient grouping and replication protocol named S-DATA, which ensures content persistence over a non-persistent P2P network for web hosting.

References

1. R. Bhagwan, D. Moore, S. Savage, and G. Voelker. Strategies for highly available peer-to-peer storage systems. In *Proc. FuDiCo*, May 2002.
2. R. Bhagwan, K. Tati, Y. Cheng, S. Savage, and G. Voelker. Total recall: system support for automated availability management. In *Proc. NSDI*, 2004.
3. S. Blond, F. Fessant, and E. Merrer. Finding good partners in availability-aware p2p networks. In *Proc. SSS*, 2009.
4. W. J. Bolosky, J. R. Douceur, D. Ely, and M. Theimer. Feasibility of a serverless distributed file system deployed on an existing set of desktop pcs. In *Proc. ACM SIGMETRICS*, 2000.
5. F. E. Bustamante and Y. Qiao. Friendships that last: Peer lifespan and its role in p2p protocols. In *Proc. Web Content Caching and Distribution*, pages 233–246, 2004.
6. J. Chu, K. Labonte, and B. N. LevineH. Availability and locality measurements of peer-to-peer file systems. In *Proc. ITCom*, 2002.
7. Y. Dan, C. XiuMeng, and C. YunLei. An improved p2p model basedon chord. In *Proc. 6th PDCAT*, 2005.
8. J. R. Douceur. Is remote host availability governed by a universal law. *Performance Evaluation Review*, 31(3):25–29, 2003.
9. K. Kim. Time-related replication for p2p storage system. In *Proc. ICN*, pages 351–356, April 2008.
10. R. Mahajan, M. Castro, and A. Rowstron. Controlling the cost of reliability in peer-to-peer overlays. In *Proc. IPTPS*, 2003.
11. J. W. Mickens and B. D. Noble. Exploiting availability prediction in distributed systems. In *Proc. NSDI*, 2006.
12. K. Ranganathan, A. Iamnitchi, and I. Foster. Improving data availability through dynamic model-driven replication in large peer-to-peer communities. In *Proc. GP2PC*, 2002.
13. K. Rzadca, A. Datta, and S. Buchegger. Replica placement in p2p storage: Complexity and game theoretic analyses. In *Proc. DCS*, pages 599–609, June 2010.
14. J. Sacha, J. Dowling, R. Cunningham, and R. Meier. Discovery of stable peers in a self-organising peer-to-peer gradient topology. In *Proc. IFIP DAIS*, 2006.
15. S. Saroiu, P. K. Gummadi, and S. Gribble. A measurement study of peer-to-peer file sharing systems. In *Proc. MMCN*, 2002.
16. T. Schwarz, Q. Xin, and E. Miller. Availability in global peer-to-peer storage systems. In *Proc. IPTPS*, 2004.
17. N. Shahriar, M. Sharmin, R. Ahmed, M. Rahman, R. Boutaba, and B. Mathieu. Diurnal availability for peer-to-peer systems. In *Proc. CCNC*, Las Vegas, Nevada, USA, Jan 2012.
18. S. Shi, G. Yang, J. Yu, Y. Wu, and D. Wang. Improving availability of p2p storage systems. In *Proc. APPT*, 2003.
19. E. Sit, A. Haeberlen, F. Dabek, B. Chun, H. Weatherspoon, R. Morris, M. F. Kaashoek, and J. Kubiatowicz. Proactive replication for data durability. In *Proc. IPTPS*, 2006.
20. D. Stutzbach and R. Rejaie. Understanding churn in peer-to-peer networks. In *Proc. IMC*, pages 189–202, 2006.
21. Z. Yang, J. Tian, and Y. Dai. Towards a more accurate availability evaluation in peer-to-peer storage systems. *Intl. Journal of High Performance Computing and Networking*, 6(3/4):233–246, 2010.

Chapter 6
Conclusion

Peer-to-peer technology has been in use for more than a decade now. The most prominent and successful applications of this technology include file sharing, content distribution and video streaming. Yet, Web hosting and Web-based multimedia content sharing, a rather potential P2P application, have not been well-investigated. In this book we have highlighted the challenges in P2P web hosting and presented our solutions for addressing these challenges.

P2P Web hosting is a challenging problem due to the fundamental differences between P2P and client-server architectures. In the traditional Web hosting scenario, content location and content hosts are fairly persistent. On the contrary, P2P networks are characterized by their volatility in peer population and frequent content relocation. This fundamental difference introduces three main challenges for P2P Web hosting: naming, searching and availability. In this book we have presented an integrated framework, named pWeb, for solving these three challenges. pWeb offers a location-independent, persistent and Web technology compatible naming scheme over non-persistent P2P networks. Searching in pWeb is done in a distributed manner, which does not require centrally controlled compute resources (e.g., datacenters). In pWeb, the participating stable peers collaboratively index web-documents and ensures distributed ranking. Finally, pWeb exploits the diurnal availability pattern, in order to group peers in small groups. Web-contents are fully replicated between the members of a group. pWeb framework ensures that at least one peer from a group is available all the time with very high probability.

Currently, we are in the process of deploying the pWeb framework for public use. The latest version of the pWeb client can be obtained from pWeb projects homepage (www.pwebproject.net). However, while deploying this framework we are facing a few technical and deployment challenges as discussed below.

The technical challenges are more implementation specific and less research intensive. The first technical challenge we are facing is related to the NAT traversal issue. In a few cases both sides of a connection are behind NAT. To allow connectivity in such situations, we are deploying STUNT servers. However, STUNT servers can become performance bottleneck and can reduce the overall scalability of the pWeb framework. The second technical challenge is related to the heterogeneity

R. Ahmed and R. Boutaba, *Collaborative Web Hosting*, SpringerBriefs in Computer Science, DOI 10.1007/978-3-319-03807-0__6, © The Author(s) 2014

in the client platform. Currently, pWeb supports Android and Windows platforms. Yet, there exists a wide variety of other platforms. It is very difficult to develop pWeb client software for all of these operating systems and ensure compatibility across these clients.

Besides the technical challenges, we have to address a deployment challenge that is related to the contemporary business model adopted by the ISPs. Traditionally, users has to pay for both download and upload bandwidth usages. This policy will discourage a user to host Web contents from his own devices. However, we argue that the ISPs should remove the charges on upload bandwidth usage for their own benefit. Our argument can be justified by observing the economic tension between the ISPs and the content providers. ISPs work as a carrier, while the content providers take the major cut from the aggregate revenue. By allowing free upload bandwidth, ISPs can play the role of a content provider and can eventually increase their profit margin.

The Web has a tremendous importance worldwide. It has arguably become the world's greatest resource for information, and it's success has fostered a variety of new ways for people to share information, communicate, and interact. Over the past decade, a wave of cultural phenomena – including Google, Wikipedia, YouTube, MySpace, Facebook, and Twitter – have all utilized the web as their interface.

Serverless web hosting can have a dramatic social impact. pWeb will allow the free hosting of websites, without limitation on content type or size. This will provide anybody the opportunity to publish to the masses, rather than restricting them by economics. In addition, freedom of speech is a valued principle, however worldwide there are many who strive to block access to certain information. The distributed approach of pWeb is inherently resistant to censorship, and will help to spread this freedom worldwide.

The "free" nature of pWeb hosting will also promote other economic activities. From a content provider's point of view, it will allow them to focus on the production and sharing of content, rather than on hosting and administration. For an individual or small organization, they will have the capability to reach a far larger audience than would have been previously possible, due to prohibitive hosting costs. In addition, we believe that this restructuring will create the opportunity for a host of new applications that better connect the content provider with the user.